William A. Kerr, PhD
Kurt K. Klein, PhD
Jill E. Hobbs, MA
Masaru Kagatsume, PhD

Marketing Beef in Japan

Pre-publication
REVIEWS,
COMMENTARIES,
EVALUATIONS . . .

"**T**he recent liberalization of the Japanese beef system has created an export market that is expected to reach $5 billion by the end of the century. This potential has created interest in Japan's beef sector among businesses, academics, and policy makers. One key to this marketplace is the history and culture that underlie Japanese thinking and Japanese business practices. This new volume provides the reader with the historical and cultural overview, and the in-depth knowledge of the Japanese beef sector that is required. THE AUTHORS ARE INTERNATIONALLY RECOGNIZED SCHOLARS IN THIS AREA AND HAVE CREATED A TEXT THAT IS WELL WRITTEN, INFORMATIVE, AND FASCINATING."

Dermot J. Hayes, PhD
Department of Economics,
Iowa State University

More pre-publication
REVIEWS, COMMENTARIES, EVALUATIONS . . .

"**T**HIS BOOK . . . PROVIDES AN INSIGHTFUL AND AC-CURATE PICTURE OF THE DI-VERSE AND COMPLEX MARKET FOR BEEF IN JAPAN TODAY, following the 1988 Beef Market Access Agreement. It includes the potential benefits and potential difficulties in developing or expanding a niche in this unique and sometimes perplexing marketplace. This book will assist students and academics in understanding the complexities of gaining entry into a portion of the Japanese market, marketing managers and companies wanting to export beef to Japan, and those exporters who already have a foothold in this market and want to expand it."

Dennis McGivern, M.Ec.
Product and Pricing Manager,
XL Meats, Calgary, Alberta

"**T**he book fills a long felt need for a comprehensive, easy-to-read compendium on the Japanese Beef Market.

THE STRENGTH OF THE TEXT LIES IN THE STRONG DESCRIPTIVE APPROACH TO THE EXPLANA-TION OF THE MARKETING INSTI-TUTIONS IN JAPAN. In the current days of excesses in econometric analysis, it is a pleasure to read a well-written text which balances the economic literature scales somewhat."

Murray Hawkins, PhD
FAIC FCAEFMS,
Director & Wesfarmers Professor,
Muresk Institute of Agriculture,
Curtin University, Australia

Food Products Press
An Imprint of The Haworth Press, Inc.
New York • London • Norwood (Australia)

Marketing Beef in Japan

FOOD PRODUCTS PRESS
Agricultural Commodity Economics,
Distribution, and Marketing
Andrew Desmond O'Rourke, PhD
Senior Editor

Marketing Beef
in Japan

William A. Kerr, PhD
Kurt K. Klein, PhD
Jill E. Hobbs, MA
Masaru Kagatsume, PhD

Food Products Press
An Imprint of The Haworth Press, Inc.
New York • London • Norwood (Australia)

Published by

Food Products Press, an imprint of The Haworth Press, Inc., 10 Alice Street, Binghamton, NY 13904-1580

Library of Congress Cataloging-in-Publication Data

Marketing beef in Japan / William A. Kerr . . . [et al.].
    p. cm.
    Includes bibliographical references and index.
    ISBN 1-56022-044-9: acid free paper.
    1. Beef–Japan–Marketing. 2. Beef industry–Japan. 3. Beef industry–United States. 4. Beef–United States–Marketing. 5. Export marketing–United States. I. Kerr, William A. (William Alexander)
HD9433.J32M37   1994
644′.92′0688–dc20

93-23224
CIP

This book is dedicated to our parents:

Marjorie and Charlie

Ruth and Oliver

Julia and Francis

Kuriko and Chuushichi

CONTENTS

List of Tables

List of Figures

Picture Plates

ABOUT THE AUTHORS

William A. Kerr, PhD, is Professor of Agricultural Economics at The University of Calgary in Canada. He has extensive research experience in the international trade of livestock and meat products, including the Canada-U.S. Free Trade Agreement, GATT negotiations, EC-Canadian trade disputes, and Pacific Rimmarkers. He is a member of the International Association of Agricultural Economists, the International Agribusiness Management Association, and the Canadian Agrimarketing Association.

Kurt K. Klein, PhD, is Professor of Economics at the University of Lethbridge, Canada, and has been a Visiting Professor at Hokkaigakuen Kitami University and Otaru Commercial University in Japan. A former beef producer with considerable production economics experience in government and universities, Dr. Klein has spent extended periods in Japan studying the cultural, economic, and technical facets of all stages of the Japanese beef system. He is a member of the International Association of Agricultural Economists and the Canadian Agricultural Economics and Farm Management Society.

Jill E. Hobbs, MA, is a Food Marketing Economist at the Scottish Agricultural College in Aberdeen. Her research and teaching interests include trade and marketing opportunities for the international agribusiness sector. In particular, her research has focused on Canadian-Japanese and EC-Japanese beef trade, the development of Eastern European food distribution systems, and food safety and biotechnology issues for agribusiness firms. Ms. Hobbs belongs to the International Agribusiness Management Association and the International Association of Agricultural Economists.

Masaru Kagatsume, PhD, is Associate Professor of Agricultural Economics at Kyoto University in Japan and Research Fellow at Lincoln College, Canterbury University, in New Zealand, and at Queensland University, Australian National University. A former senior research staff member in the National Research Institute of Agricultural Economics, Ministry of Agriculture, Forestry and Fisheries, Dr. Kagatsume has spent extended periods in Australia and New Zealand conducting research on agricultural trade between Australia, New Zealand, and Japan. He is a member of the International Association of Agricultural Economists and the Japanese Agricultural Farm Management Society.

PART I.
INTRODUCTION

Chapter 1

Beef in Japan–The Market of Opportunity

The Japanese market for beef is unique among the world's beef markets. In many ways it is an enigma, presenting questions with few answers. This is true for those outside Japan who would like to supply beef to this growing market as well as the Japanese themselves. Beef prices in Japan are notoriously high–why is this the case and what effect has it had on the consumption of beef? Japanese tastes for beef are very different from those in Western countries–why is this and what are the ramifications for the international marketing strategies of foreign beef exporters? What other factors have influenced Japanese beef consumption in the past and how are these factors expected to affect beef consumption in the future? These and many other questions are answered in this volume. In addition, an attempt is made to provide insights that will aid in the formulation of strategies for marketing beef in Japan.

The success of any international marketing strategy will be dependent, to a large extent, on the degree to which the international marketer understands the target market. The first step in understanding any market must be the identification of those product characteristics that are desired by consumers. This requires an understanding of the forces that shape consumer purchases. The many unique facets of the Japanese market for beef make it an excellent case study with which to illustrate the complexity often associated with international marketing. A brief review of past Japanese beef consumption trends and an explanation of why this market represents new opportunities for those interested in international beef marketing are the points of embarkation for this in-depth investigation of Japanese beef consumption.

PAST TRENDS IN JAPANESE BEEF CONSUMPTION

Beef consumption does not have a long history in Japan. Powerful religious and political influences have affected Japanese food consumption patterns for many centuries. One result is that beef has only recently come to have a significant place in the Japanese diet. For over 1,000 years prior to the Meiji Restoration in 1868, the consumption of beef was illegal due to a dietary ban on the eating of flesh from four-legged animals (Yoshida and Klein, 1990).

Even after the dietary ban was lifted, however, beef consumption did not increase appreciably for almost 100 years. The religious influences that had led to the ban on meat consumption still pervaded much of Japanese society. As a result, the Japanese diet remained largely based on rice, soybeans, and fish. The fact that Japan is an island nation helped maintain the tradition of relying on the sea as the main source of protein in the Japanese diet.

Up until the second World War, consumption patterns changed very little. Since the war, however, there have been many changes in the Japanese diet, including a gradual increase in meat consumption. The Japanese diet has been progressively diversifying. Despite the traditional reliance on fish as the sole meat product in the Japanese diet, consumption of chicken and pork in particular–but also beef–has increased. The reasons for these changes will be discussed in later chapters; they include rising incomes, greater exposure to Western cooking, and greater availability of domestically produced livestock.

To give an indication of the magnitude of these changes, between 1962 and 1986 per capita beef consumption in Japan grew by 200 percent, compared to a 19 percent increase in per capita fish consumption (Hobbs and Kerr, 1990). Increases in chicken consumption (900 percent) and pork consumption (275 percent) over this period were even greater than that for beef.

There exists a wide band of beef prices in Japan, reflecting the very wide range in the quality of beef consumed. The best quality beef according to the Japanese, which commands extremely high prices, is well-marbled. "Marbling" describes the degree of intramuscular fat in meat. Extremely heavily marbled beef has a frost-like appearance, as it is entirely speckled with fat. This beef is best suited for traditional Japanese dishes, which require the beef to be cut into wafer-thin slices that are then boiled. Hence, while beef has emerged as part of the modern Japanese diet, the dishes consumed are unique to Japan. Highly marbled beef is considered to be a delicacy. In contrast, what is considered the highest quality beef in most Western countries is a lean product.

It will be important, therefore, for an off-shore beef producer who is contemplating supplying the Japanese market to understand the very short history, and hence the lack of tradition, which characterizes beef consumption in Japan. Understanding the development of Japanese beef preferences may enable off-shore producers to tailor their beef product to meet this very unique Japanese demand. If this can be done, there may be an opportunity to earn considerable profits given the high market prices these products command.

If substantial profit opportunities exist in the Japanese beef market, then why have off-shore beef producers not exported large amounts of well-marbled beef to Japan in the past? The main reason has been the import regulations restricting both the amount and quality of beef that could be exported to Japan.

A REVIEW OF THE PREVIOUS BEEF-IMPORTING SYSTEMS

Japan has one of the world's most highly protected agricultural sectors, and the beef industry is no exception. The rate of protection afforded the beef industry is ranked third after only the rice and dairy industries.

This highly protectionist policy has arisen from a combination of social and political factors. The latter results from the strong influence that agricultural interests have over Japanese politics. The Liberal Democratic Party (LDP) has been in power in Japan continuously since 1955, and is heavily dependent on support from rural communities. Furthermore, although the Japanese population has become highly urbanized over recent decades, urban families maintain strong ties to their rural heritage. Many urban Japanese still operate or own small areas of land. Despite migration to the cities, the Diet (Japanese parliament) has never been fully restructured to reflect this demographic change, and rural communities control a disproportionately large number of seats.

Farmers' interests are effectively voiced through the agricultural cooperatives, or Nokyo. The influence of the Nokyo pervades almost every sector of Japanese agricultural life, from village-level to national level representation. These cooperatives lobby the government in support of agricultural interests. Furthermore, many members of the LDP began their political careers within the cooperatives and therefore owe much of their success to the support of these cooperatives.

Thus, agricultural interests in Japan have substantial political leverage, a leverage that extends far beyond the economic significance of the agricultural sector in the economy as a whole. This provides one of the reasons for the intensely protectionist agricultural policies that have been followed by Japanese governments.

A second reason for agricultural protectionism has been traditional Japanese concerns over food security. These concerns were heightened by the experiences of post-World War II food shortages, as well as by the extension of coastal fishing areas by most nations to 200 miles (320 kilometers) and the American soybean embargo of 1973. Japan's large population is supported by a limited agricultural resource base—much of Japan's terrain is mountainous and, therefore, unsuitable for most forms of agriculture. Food self-sufficiency has been a major priority in Japan. Their self-sufficiency ratios, however, have fallen very rapidly in the last 30 years, from about 90 percent in 1960 to 71 percent in 1990 (Hayami and Yamada, 1991).

Furthermore, Japanese agriculture is typically small-scale and high cost. In 1990 the average beef herd size was approximately 12 head. In the past if there was to be a domestic beef industry contributing to self-sufficiency, these small, inefficient domestic beef producers had to be protected from the competition provided by foreign imports. In addition, the growing disparity between urban and rural incomes during the late 1950s and early 1960s led to the use of protectionist measures in an attempt to maintain or enhance income levels in rural areas.

For beef exporters, the first indication of this protectionist attitude came in 1954 when a 10 percent import tariff was imposed. In 1958 the tariff was supplemented with an import quota based on the value of beef imports. In addition, importing firms in Japan needed a license and an allocation of foreign exchange before they were able to import beef.

In 1961, the Price Stabilization Law for Livestock Products provided for the creation of the Livestock Industry Promotion Corporation (LIPC). This quasi-governmental organization became responsible for the administration of those government policies applied to the livestock industry. In 1966, the LIPC was brought specifically into the beef importing system when it was given the role of overseeing the price stabilization scheme for beef products. The beef import tariff was raised to 25 percent in 1964, and import quotas based on physical quantities were introduced, replacing the quotas based on the value of imports.

The LIPC was empowered to buy, sell, and hold stocks of imported beef so that it could carry out its mandate to stabilize domestic prices. By 1988, the LIPC controlled approximately 80 percent of Japan's beef imports. Figure 1.1 shows the allocation of beef import quotas in Japanese Fiscal Year (JFY) 1987.[1]

The global quota was divided between a general quota (90 percent) and special quotas (10 percent). The special quotas catered to specific markets and, in most cases, represented different qualities of beef than did the general quota segment.

The general quota was divided into a private quota–which was administered by the Japan Meat Conference (JMC)–and the LIPC quota. The JMC collected levies on the sale of its quota beef, as did the LIPC.[2]

The "Merchant Tender System" was the primary mechanism used by the LIPC to manage beef imports. Eighty percent of LIPC beef imports were obtained through this system. The LIPC normally specified very narrow quality requirements for the beef imported under this system. Imports were primarily restricted to lean

FIGURE 1.1. Division of Japanese beef import quotas (Japanese fiscal year 1987).

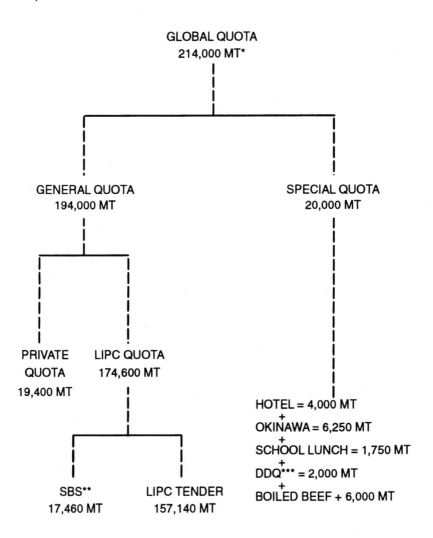

* = Metric Tons
** = Simultaneous-Buy-Sell system
*** = Demand Development Quota

frozen beef. Specifying imports of this quality provided two important advantages to the LIPC. First, since the beef was in frozen form, the LIPC could store it, thereby manipulating the stocks of frozen imported beef to stabilize domestic beef prices. Second, and perhaps more significantly, Japanese consumers prefer well-marbled chilled beef. Hence, imported beef did not exhibit the characteristics most preferred by Japanese consumers. Lean frozen beef is seen as inferior in quality to well-marbled, chilled beef.[3] Thus, the domestic beef industry could be protected from competition with imports because consumers would be unlikely to substitute this "inferior" beef for high quality domestic beef.

Using the Merchant Tender System, the LIPC allocated quotas for intermediate or end-users, such as associations of meat canners or consumer cooperatives. These users then contracted with an importing firm to arrange for the importation of their beef. The Ministry of Trade and Industry (MITI) specified which firms were allowed to import beef–there were approximately 36 of these firms (Longworth, 1983). As control of beef imports passed through the various stages, each stage attempted to extract profits or quota rents from its control of imports. Thus, the 36 importing firms tried to maximize the difference between the landed import price and the price at which they sold the beef to the quota holders. The LIPC was able to extract quota rents and levies from the issuance of quotas. These funds were then used to modernize the domestic beef industry.

Foreign producers wishing to export beef to Japan had to be on a Preferred Brand List. The product and packaging had to be of a consistent quality and had to satisfy standards set by the Japanese buyers. Thus, it was extremely difficult for exporting firms to "learn by experience." If a foreign exporter was removed from the Preferred Brand List, it was extremely difficult to get back onto that list. Quantity, quality, and participation in the Japanese beef importing system were, therefore, controlled by the LIPC.

The LIPC used three other methods to distribute smaller portions of its quota. The One-Touch quota allocation system, introduced in 1970, enabled authorized importers to sell directly to designated distributors without involving the LIPC as an intermediary. The LIPC did, however, supervise the transaction and collect a levy on the sale.

In 1978, a Modified Tender System was introduced and, for the first time, chilled beef could be imported. The LIPC purchased beef from the importing firm that tendered the lowest price and then re-sold the beef to the end-user who offered the highest price (Lloyd, Frank, and Faminow, 1987).

In 1985, the Simultaneous-Buy-Sell system (SBS) was introduced. Foreign suppliers and Japanese importers simultaneously submitted bids. Quotas were then allocated in descending order to the foreign suppliers with the lowest bids and to the Japanese importers with the highest bids. The LIPC collected the difference between these two prices. Since the LIPC did not specify the quality of beef to be imported, this system heralded a change in Japanese beef-importing policy. However, prior to 1988, the SBS system applied to frozen beef only. Thus, the Japanese government was still providing protection to the domestic beef industry by restricting beef imports to beef whose quality did not conform to consumer preferences.

Furthermore, the position of the LIPC as an intermediary in almost all sales of imported beef prevented beef exporters from building close ties with Japanese buyers. The LIPC acted as a buffer in the marketing chain, inhibiting the normal relationship between buyer and seller.[4]

The Japanese government utilized its beef price stabilization scheme to further regulate the beef market. This scheme had pre-set price bands for domestic beef prices in Japan. The LIPC was given the task of keeping domestic prices within these bands. It could strategically buy and sell stocks of imported beef–and later, domestic beef–to keep prices within these bands.[5]

The restrictions on the quantity and quality of imports, as well as the number and types of firms that could participate in the importing system, caused much international discontent. Pressure to liberalize the beef-importing system grew throughout the 1970s and 1980s. The U.S., in particular, exerted considerable political pressure on Japan, resulting in a series of agreements that gradually eased Japanese import restrictions and provided new opportunities for those wishing to export to the Japanese market.

In 1978, the Strauss-Ushiba agreement between the U.S. and Japan established a High-Quality Beef (HQB) portion of the general quota. This was a specific quota for the lean, grain-fed beef typically produced by the U.S. beef industry. The Brock-Yamamura Agreement between the U.S. and Japan followed in August 1984 and led to an increase in the HQB quota.

Political pressure on Japan continued throughout the 1980s and culminated in the signing of the Beef Market Access Agreement (BMAA) between the U.S. and Japan on June 20, 1988. A similar agreement was signed between Australia and Japan later that summer.

TERMS OF THE 1988
BEEF MARKET ACCESS AGREEMENT

The Beef Market Access Agreement (BMAA) has led to a major liberalization of the Japanese beef-importing system and has been greeted with much enthusiasm and optimism by major beef-exporting nations.

The BMAA introduced three significant changes to the Japanese beef-importing system. The first change was the removal of all beef import quotas. Quotas were gradually expanded during a transitional phase, from a level of 214,000 metric tons (mt) in Japanese Fiscal Year (JFY) 1987–rising by 60,000 mt per year–to a final level of 394,000 mt in JFY 1990. After this date, quotas were completely removed. An increased import tariff was imposed in place of the quotas. In JFY 1991, the year in which all quota restrictions were finally removed, the import tariff rose from 25 percent to 70 percent, and then fell to 60 percent in JFY 1992 and to 50 percent in JFY 1993. The tariff is bound at 50 percent[6], and any further reductions are subject to negotiation at the General Agreement on Tariffs and Trade (GATT) talks. Table 1.1 summarizes the main details of the BMAA.

An emergency provision allows the tariff to be increased by 20 percent if beef imports increase by more than 20 percent of the previous year's level. The emergency provision is in effect for the post-transitional phase from JFY 1990 to JFY 1993. Table 1.1 indicates the maximum import levels that would trigger this emergency provision.

The second major change to emerge from the BMAA is the liberalization of participation in the Japanese beef market. As Table 1.1 indicates, the LIPC share of the quota was steadily reduced

TABLE 1.1. The U.S.-Japan Beef Market Access Agreement – phase-in provisions.

	JFY 1987	JFY 1988	JFY 1989	JFY 1990	JFY 1991	JFY 1992	JFY 1993
GLOBAL QUOTA*	214,000	174,000	334,000	394,000	0	0	0
Special Quotas	20,000	25,000	27,000	30,000	0	0	0
General Quota	194,000	249,000	307,000	364,000	0	0	0
LIPC Quota	174,600	224,100	276,300	327,600	0	0	0
Simultaneous-Buy-Sell	17,460	67,230	124,335	196,560	0	0	0
Participants	17,460	22,410	27,630	32,760	0	0	0
Any Importer	0	22,410	48,353	81,900	0	0	0
Any End-User	0	22,410	48,353	81,900	0	0	0
LIPC Tender	157,140	156,870	151,966	131,040	0	0	0
Private Quota	19,400	24,900	30,700	36,400	0	0	0
Tariff	25%	25%	25%	25%	70%	60%	50%
LIPC Surcharge	yes	yes	yes	yes	no	no	no
Trigger Level of Imports for Invocation of Emergency Measures	N/A**	N/A	N/A	N/A	yes	yes	yes
Tariff if Emergency Measure Invoked					90%	80%	70%

*All Quantities in Metric Tons
**Not Applicable

Source: Beef Market Access Agreement

during the transitional phase, while the SBS quota was increased. Previously, the SBS quota segment had been restricted to LIPC designated "participants." The transitional phase, however, provided for two new SBS user categories: "any importer" and "any end-user." The greater part of the quota growth took place in these two categories.

The third major change brought about by the BMAA was the removal of the LIPC from direct involvement in the beef-importing system as of April 1, 1991.

PRACTICAL RAMIFICATIONS OF THE BEEF MARKET ACCESS AGREEMENT

The removal of import quotas means that foreign beef producers have far greater access to the Japanese beef market. Furthermore, import business is no longer restricted to the 36 designated importing companies. The removal of the LIPC allows much closer ties between exporter and end-user. This means that the preferences of Japanese consumers should begin to shape the types of beef that are imported.

The removal of the LIPC has had important ramifications for the types of beef imported. The LIPC tendering system had allowed imports to be specified for lean frozen beef only. As a result of the BMAA, it is now possible to supply the Japanese mass market with the chilled, well-marbled beef it prefers.

One of the unique characteristics of the Japanese beef market is the wide range of beef products it demands, from highly marbled beef to cheaper lean beef. Thus, there will still be a market niche for lean frozen beef. However, the high prices that offer the potential for large rewards apply to chilled, well-marbled beef. The BMAA provides foreign beef producers with an opportunity to compete for these high-priced market segments if they so choose.

Competing in this newly liberalized market, however, is likely to pose a significant challenge to foreign beef producers. The production of well-marbled beef is poorly understood outside of Japan. Changes to production practices will be essential; cattle must be fed for longer periods of time to achieve heavy marbling. Different breeds of cattle may have to be used or developed. If chilled beef is to be exported to Japan then it may require a considerably longer shelf-life than that required for chilled beef in an exporter's domestic market. This may be particularly important if air shipment of beef becomes prohibitively expensive with the new, high *ad valorem* tariff in place.[7] Therefore, exporters may need to invest resources in improving the shelf-life of their beef.

SUMMARY

The previous Japanese beef-importing system forced exporters to sell the beef product that they produced for their own markets but in frozen form. With the LIPC in place, importers and exporters were not able to deal directly with one another and consumer preferences did not determine the types of beef imported. The provisions of the BMAA have forced beef exporters to become more cognizant of consumer preferences in Japan. Exporting beef to Japan entails considerable risks. Exporters are no longer able to simply export a product with the characteristics that satisfy their domestic consumers. They will have to establish relationships with Japanese importers. Thus, many unknowns now face the would-be exporter of beef to Japan.

Perhaps the most crucial task beef exporters must undertake is to understand the nature of beef consumption in Japan. With an understanding of the factors affecting the demand for beef, an international marketing strategy can be developed and the exporter's product can be aimed at the most appropriate segment of the Japanese market.

The analysis of consumer habits can be approached from a number of directions. The approach chosen for this book is that used by economists and is generally termed *demand analysis*. Demand analysis provides the means for systematically identifying the major forces that affect consumption and allows for the discussion of markets as well as individuals in both the short and long term. A brief review of the basics of demand theory is presented in the Appendix at the end of the book. This Appendix provides a framework for the detailed analysis found in the remainder of the book. Those who are well acquainted with demand relationships may proceed directly to Chapter 2.

PART II.
FACTORS THAT SHIFT DEMAND

Changes in a number of factors can be expected to alter the quantity of beef consumed in Japan–to shift the demand for beef. The ways in which these factors interact with consumer preferences will determine the growth (or decline) of the market for beef in Japan. In Part II these relationships are examined in depth. Chapter 2 examines the effects of both past and current changes in Japanese consumer tastes. In Chapters 3 and 4 the effects of changing demographics and rising incomes are analyzed in turn. Important government regulations are examined in Chapter 5. Market trends for the major beef substitutes–pork, chicken, seafood, and soya products–are the subject of Chapter 6.

Chapter 2

Beef and Japanese Tastes

Japanese consumers have unique beef preferences when judged against the tastes of most Western consumers. The taste difference has three major manifestations. First, Japanese tastes in beef fall along a much wider quality spectrum than in Western countries. This is reflected, to a certain extent, in the very wide range of beef prices that can be found in a Japanese supermarket. Kerr and Klein (1989) reported that prices can range anywhere from $4 per kilogram to $450 per kilogram in Japanese supermarkets. Hence, the beef market is very segmented, with many quality-based niches.

The second way in which Japanese beef tastes are unique is the characteristic that is most rewarded in the market. In Japan, the highest quality beef is that which is heavily marbled. By contrast, beef considered to be of the highest quality in most Western industrialized countries is much leaner.

Finally, Japanese people also consume much smaller amounts of beef than do consumers in other industrialized countries. In 1986, the average per capita consumption of beef in Japan was about 6 kilograms (kg). This compares to about 40 kg per capita in Canada and 50 kg per capita in the U.S. (OECD, 1986).

Development of these unique beef tastes is many-faceted and includes the influences of religion, history, politics, law, and culture. These influences, which are fundamental to an understanding of the Japanese beef consumer, are the subject of this chapter.

RELIGION, HISTORY, AND CULTURE

Religion

Religious influences have played an important part in the development of the Japanese diet. Buddhism and Shintoism have been, and remain, an important part of Japanese life.

Buddhism was introduced into Japan early in the sixth century A.D. One of the central beliefs of Buddhism is that all beings–humans, animals, and insects–pass through a cycle of life, death, and rebirth. Rebirth occurs in one of six "realms," ranging from heaven to hell. The achievement of "Buddhahood" is the only way an individual can escape this cycle (Yoshida and Klein, 1990). Being born into a more favorable or higher realm depends on the actions of a being during its current lifetime. To be reborn into a higher stage of life an individual must commit only good acts during its current life stage. Committing evil acts condemns an individual to be reborn as a lower form of life.

According to Buddhist doctrine, the killing of another being–be it human, animal, or insect–is the worst form of evil. Thus, with the spread of Buddhism came an aversion to all killing, including the killing of animals. This was already manifest during the Heian Period (782-1181) when, despite many famines, "cattle and horses were laying in the field and skeletons were seen in the street" (Yoshida and Klein, 1990, p. 135).

Over time, the influence of Buddhism began to pervade all aspects of domestic life. During the Edo period (1596-1868) the Tokugawa Shogunate decreed that every family should register with a certain Buddhist sect. By this time, it became commonplace for homes to have a Butsudan–a small cabinet containing an image of Buddha–to which offerings of food, flowers, and incense were often made.

The second main religious influence was Shintoism, the indigenous Japanese religion. Shintoism teaches that involvement with death pollutes an individual. This became linked with the Buddhist idea of not killing animals.

The joint influences of Buddhism and Shintoism meant that the Japanese grew to detest the sight of blood and the dead bodies of people and animals (Simpson et al., 1985). While the emergence of nuclear families has meant a lessening in the influence of religious practices, their effect on the development of the traditional Japanese diet should not be overlooked. Many facets of the traditional Japanese diet, with its restrictions on the consumption of animal flesh, still exert a strong influence on modern-day consumer preferences.

History

Japanese history records that the early ancestors of the present Japanese people ate beef during the Yamata period (A.D. 200 - A.D. 644). However, during the Asuka and Nara Periods (645-781) the Taiho-Ritsuryo Law was passed. This law forbade the eating of meat from four-legged animals and was a direct result of the growing influence of Buddhism. This law remained in place until the Meiji Restoration in 1868 (Yoshida and Klein, 1990).

Over the centuries, starting in the Asuka and Nara Periods, various edicts reinforced the prohibition of meat consumption. For example, farmers in the Kinki region of Japan were found to have disobeyed the prohibition and consumed beef on special occasions. Their practices were then specifically banned during the Heian Period (782-1181). Between 1685 and 1709, the Shogun Tokugawa Tsunayoshi issued several edicts that forbade cruelty to animals. The edicts became increasingly extreme, and eventually the trapping and killing of all birds and animals were banned (Yoshida and Klein, 1990).

With the Meiji Restoration, the laws and edicts prohibiting the killing of animals and the eating of meat were rescinded. The young Emperor Meiji appeared anxious to encourage the Westernization of some aspects of Japanese society. In 1871, the Meiji government announced that "a meat diet is superior to keep nourishment and health." The following year, Emperor Meiji suggested that his citizens eat beef, and he personally ate meat as a way to encourage the Westernization of his country.

The eating of meat from four-legged animals has been legal since 1868. However, one must still consider the inevitable impact on Japanese culture, values, and tastes of a meat-eating ban that existed for over one thousand years.

Culture

Japanese society has traditionally been characterized by values of collectivism rather than individualism. Thus the group, and in particular the family, is a very important influence in Japanese culture. The Edo Period (1596-1868) saw the introduction of the Confucian ideology. Confucianism taught correct observance of social and hierarchical relationships, providing for a rigidly controlled social system. It also taught a strict respect for the law. Thus the social values inherent in a Confucian society reinforced the law banning meat consumption.

Farm animals, draught animals, and horses for the military, of course, continued to be an important component of the Japanese economy. Hence, diseased animals, and those that died of natural causes, still had to be disposed of. The combination of the legal ban on consuming meat from four-legged animals and the moral influences of Buddhism and Shintoism led to discrimination against people involved in these activities. These people became known as the Eta class. They were considered to be polluted by their association with the death of an animal, and this pollution was thought to be contagious. The Eta class performed all butchery and leather-crafting tasks in Japanese society. The Meiji government officially abolished discrimination against the Eta class in 1871.

The centuries-long discrimination against this class of people demonstrates the depth of Japanese cultural feelings about the killing of animals. Thus, the consumption of beef in Japan has an entirely different cultural origin when compared to those factors which have influenced the consumption of beef in Western industrialized countries.

CULINARY ARTS

Traditional Japanese beef dishes are very different from beef dishes consumed in Western countries. This arises largely from the characteristics of the beef consumed, which is partially a direct result of religious, historical, and cultural factors.

The aversion to death, killing of animals, and blood by more traditional Japanese consumers means that their meat must be cut in a form that does not show blood. Hence, Japanese butchers cut meat into wafer-thin slices. Furthermore, much of the meat is cooked by dipping it into boiling water because Shintoism teaches that water is a means of purification (Simpson et al., 1985).

For wafer-thin slices of beef to be successfully cooked this way, the meat must have a high degree of intramuscular fat, or marbling. When boiled in water or oil, the fat within the muscle melts around the meat, sealing in its natural juices and keeping the meat tender (Simpson et al., 1985). When lean Western-style beef is cooked in this manner it becomes tough and leathery. Therefore, to most Japanese consumers, the more marbled the beef, the tastier the dish (Yoshida and Klein, 1990).

The extremely heavily marbled cuts of beef are often referred to as *Shimofuri niku*, or frost-like beef. The use of this type of beef, combined with the boiling method of cooking, created beef dishes unique to Japan. Nabemono dishes are the most well known of these and include Sukiyaki and Shabu-Shabu. These dishes are served in a pot and require the beef to be boiled in oil or water at or near the table.

Japanese beef dishes have taken on more variety over time. Nimono, a dish in which wafer-thin slices of beef are boiled and then used as seasoning for soups and noodles, uses the lower quality cuts of meat from the highly marbled carcasses (Simpson et al., 1985). After the Second World War, Yakiniku, or Korean barbecue, was introduced. This dish contains small pieces of beef that have been heavily seasoned and then grilled or barbecued. In the latter part of the 1960s, grills and barbecues began to appear in some Japanese homes. This increased the popularity of hamburger-type dishes, as well as stews and curries.

A wider variety of pork and chicken cooking methods, which preceded the diversification of beef cooking methods, influenced beef cuisine. Unlike beef, pork consumption became relatively popular following the Meiji restoration. Traditional Japanese pork dishes were used as a side dish to complement boiled rice. Other pork dishes were more Western in style, with broiled, roasted, or fried meat (Simpson et al., 1985).

Less chicken than pork was consumed and, initially, chicken was boiled in a manner similar to Sukiyaki and Shabu-Shabu dishes. The presence of the U.S. occupation army after World War Two introduced roast or broiled chicken to the Japanese diet (Simpson et al., 1985).

Some of these new chicken and pork cooking methods were adapted to the preparation of beef. The result of the diversity in cooking methods is that the retail cuts of beef available in Japanese stores range from wafer-thin slices to Yakiniku pieces to cube cuts to steak cuts (Simpson et al., 1985).

Domestic dairy steer beef and imported grass-fed beef have been gaining an ever-increasing proportion of the Japanese beef market. This is significant as these types of beef require considerably different cooking methods. Grass-fed beef is considered to be unsuitable for Sukiyaki or Shabu-Shabu since the meal is traditionally cooked at the table and grass-fed beef gives off odors which are offensive to the Japanese. This is because grass-fed beef has less inter- and intramuscular fat than does well-marbled grain-fed beef. As a result, the muscle protein overheats in the boiling liquid and causes the amino acids in the muscle protein to break down, producing a sulphurous odor. Well-marbled grain-fed beef, on the other hand, has sufficient fat to prevent amino acid breakdown and avoids the release of these unpleasant odors (Longworth, 1983). Australian and New Zealand grass-fed chilled beef is, however, suitable for barbecuing in the Yakiniku style as there is less risk that offensive odors will be given off during cooking. The Japanese take great care with the presentation of their food. Thus, not only is the taste of food important, so is the way in which it looks and smells (Kerr and Klein, 1989).

Beef is seldom cooked in the form of large roasts or steaks as it is in Western homes. Instead, beef is consumed in the form of roasts and steaks in Western-style restaurants. Roasts or steaks are usually "well done" since many Japanese still prefer not to eat their meat in a form that shows blood (Longworth, 1983). However, this does not mean that the Japanese like large char-grilled steaks. Many find this type of food unappetizing and are not used to the excessive chewing steaks require. In fact, traditional Japanese food, such as rice, noodles, and seafood, is soft in texture. Hence, meat should ideally be firm but tender enough to not need a great deal of chewing since the mastication of food is generally considered unsavory (Kerr and Klein, 1989).

An important part of the Japanese culinary style is the use of chopsticks. The practice of using chopsticks began in the Asuka and Nara periods (645-781). Chopstick use requires that food be cut into small pieces. Obviously, Western-style steaks and roasts cannot be eaten with chopsticks. Furthermore, few Japanese kitchens are equipped to handle the cooking or storage of beef in Western cuts or quantities.

Table 2.1 illustrates the wide range of beef dishes that comprise the Japanese diet and indicates which types of beef are used in different dishes. Only Wagyu beef is used in the preparation of traditional delicacies such as high quality Sukiyaki. The table also indicates that only Wagyu beef achieves the high Japanese grades.[1]

DIET TRENDS

General Trends

Prior to the Second World War, the Japanese obtained more than 60 percent of their total energy intake from rice. Less than one percent of their energy intake came from meat. Rice was, and to some extent still is, the staple food of the Japanese population. It

TABLE 2.1. Japanese beef dishes and types of beef used.

Dish	Beef Grade	Beef Source
Sukiyaki (High Quality)	A5, A4	Wagyu Heifer Wagyu Steer
Sukiyaki (Medium Quality)	A3, B5, B4, B3	Wagyu Heifer Wagyu Steer Imported Grain-Fed
Barbecue	A3, A2, B3, B2	Wagyu Steer Dairy Steer Dairy Heifer Imported Grain-Fed
Hamburger	A2, A1, B3, B2, C3, C2, C1	Wagyu Steer Dairy Steer Dairy Heifer Cull Dairy Cows Imported Grain-Fed Imported Grass-Fed

was either eaten by itself or with a small salty side dish (Simpson et al., 1985).

Following Japan's rapid economic growth of the 1960s, the Japanese diet began to change. More animal protein and less rice was consumed. Simpson et al. (1985) identified four main changes: (1) decreased rice consumption and increased consumption of flour-based foods such as bread and pasta; (2) increased consumption of livestock products; (3) increased consumption of fruits and vegetables in the form of desserts and salads; (4) substitution of sugar, oil, and mayonnaise for traditional seasonings such as miso[2] and soya sauce.

These changes have been called a "Japanization" of Western culinary styles rather than a "Westernization" of the Japanese diet because Western foods have been incorporated into Japanese cuisine. Thus, Japanese-style cooking and presentation have not changed substantially–rather the proportional contents of the ingredients have changed. Popular beef dishes are still made with wafer-thin beef, the only difference is that people tend to eat more slices in one meal than before (Longworth, 1983).

The changes in the Japanese diet have dramatically increased the amount of fat in the Japanese diet. Protein intake has also increased, but caloric intake has not changed a great deal. These changes were brought about by a number of factors. First, the per capita disposable income increased. Second, the reliance on Western food aid after World War Two allowed people to become familiar with wheat- and corn-based foods. Third, the School Lunch program, initiated by the Occupation Authorities, allowed children to become accustomed to new staple foods. The expansion of the urban population at the expense of the rural population, the increase in the number of nuclear families, and the increase in the number of working housewives have all contributed to the diversification of the Japanese diet (Simpson et al., 1985).

The high growth rate of GNP during the 1960s led to increased expenditure on poultry and pork products. Beef, with its high cost of production, was still too expensive for many people. The mechanization of agriculture reduced the number of beef animals kept on farms (previously these animals had been used for draught purposes). This caused a decrease in the supply of beef and a resultant rise in its price. Further, supply shortages were caused by the increasingly restrictive beef import regulations. Finally, the complex and relatively inefficient Japanese distribution system contributed to the high beef prices in Japan.

As a result, consumers substituted away from beef consumption and towards pork and poultry consumption (Simpson et al., 1985). Beef prices rose substantially between 1960 and the 1980s. Longworth (1983) reported that in 1960 beef prices were the about same or even a little lower than pork and chicken prices. However, by 1983, average beef prices were three times more expensive than average chicken prices.

The 1960s saw an increase in the consumption of livestock products. These were largely imitation products, however, such as pressed ham (an imitation ham which was made from small pressed pieces of seasoned imported mutton or horsemeat). By the late 1960s, consumers began to demand pure livestock products such as true ham from pork, natural cheese, and beef rather than chicken or pork in processed meats (Simpson et al., 1985).

The slowing of the Japanese economy after the 1973-1974 oil crisis coincided with further changes in the Japanese diet. Caloric intake peaked and became stagnant. The consumption of fruits, vegetables, flour products, eggs, and marine products peaked. Consumption of meat and milk products continued to increase, but at a slower rate after the 1978-1979 oil crisis (Simpson et al., 1985). Although the most important animal-based protein is still fish, the growth in livestock consumption has out-paced the growth in fish consumption. Over the past 30 years, pork consumption has increased by 300 percent, chicken consumption by 1,000 percent, beef consumption by 270 percent and fish consumption by 22 percent (Japan MAFF, 1992). Table 2.2 presents the average annual per capita consumption of beef, pork, chicken meat, and fish in Japan over the past 30 years.

The results of a Japanese consumer survey suggested that higher income consumers expressed a stronger preference for beef than did lower income Japanese consumers (Hayes, 1990b). The preferences of consumers from five income brackets are reported in Table 2.3. The table indicates that pork is the preferred meat in all income brackets except for the one in which annual income is greater than

TABLE 2.2. Average annual per capita consumption of beef, pork, chicken meat, and fish in Japan (carcass weight).

Year	Beef (kg)	Pork (kg)	Chicken Meat (kg)	Fish (kg)
1962	2.0	4.2	1.3	30.4
1963	2.5	3.6	1.5	30.5
1964	2.9	3.8	1.8	25.7
1965	2.9	5.1	2.1	29.8
1966	2.1	7.0	2.5	29.6
1967	2.1	7.4	3.0	31.2
1968	2.3	7.3	3.3	33.1
1969	3.1	7.6	4.0	31.4
1970	3.4	7.6	4.8	33.0
1971	3.7	8.7	5.3	33.2
1972	4.1	9.4	6.0	34.3
1973	3.9	10.7	6.5	32.3
1974	4.0	11.0	6.7	35.5
1975	4.0	11.1	6.7	35.6
1976	4.0	11.3	7.6	35.6
1977	4.4	11.9	8.2	34.5
1978	5.0	12.8	9.2	36.0
1979	5.2	14.3	9.9	34.8
1980	5.2	14.4	10.1	35.2
1981	5.7	14.2	10.3	35.4
1982	5.8	14.1	10.8	34.1
1983	6.0	14.1	11.1	34.0
1984	6.4	14.2	11.6	35.5
1985	6.6	15.0	12.0	35.8
1986	6.8	15.3	12.9	36.4
1987	7.2	16.2	13.7	36.7
1988	7.3	16.6	14.2	37.0
1989	8.0	16.7	14.3	37.0
1990	8.7	16.7	14.2	37.1
1991	9.1	16.7	14.2	—
1992 (prelim.)	9.4	16.7	14.3	—

Source: Japan MAFF, 1992.

TABLE 2.3. Consumer preferences for different meat products, by income (percentage of respondents).

Annual Income	Beef	Pork	Poultry	Other/N.A.*
>Y10,020,000	46.6	32.8	16.1	4.6
Y7,014,000-Y10,019,999	37.6	45.4	12.8	4.1
Y5,010,000-Y7,013,999	30.9	49.7	14.7	4.8
Y3,006,000-Y5,009,999	30.4	46.5	18.5	4.7
<Y3,006,000	29.5	40.5	26.0	4.1

*Not Available
Source: Japan Meat Service and Information Center 1988.

Y10,020,000. There beef is the most preferred meat. In fact, pork is the type of meat most often purchased by Japanese consumers (Khan et al., 1990). Preference for beef increases as income increases. Furthermore, only in the two highest income groups do preferences for beef change substantially. This supports the hypothesis that beef is often considered to be a luxury food in Japan.

Chicken is most preferred by the lowest income group and preference for chicken appears to decline as income increases. In the past chicken prices have been far lower than beef prices. Finally, pork preferences are the strongest in the middle income bracket (from Y5,010,000 to Y7,013,999).

Another aspect of Japanese consumer tastes is the preference for domestic beef over imported beef. This is particularly true if the imported beef is in a frozen form, as most imports have been until recently. The Japanese consumer places great value on the freshness of a product; hence frozen beef is perceived as inferior to chilled beef.

The Japanese preference for beef produced in Japan has been documented by consumer surveys. Domestic Wagyu and dairy beef are the types of beef purchased most often by Japanese consumers (Khan et al., 1990). Domestic beef is purchased in a chilled form, and product freshness is one of the main determinants of the purchase decision. It seems clear that, given a reasonably priced choice, Japanese consumers will purchase chilled rather than frozen beef. Imported beef is usually purchased because of its relatively low price rather than its taste and texture characteristics.

The preference for domestic food products even affects chicken consumption. Despite the fact that much of Japan's breeding stock comes from the U.S., Khan et al. (1990) report that many Japanese people consider domestically produced poultry to taste better than imported poultry. Hence, there appears to be a definite perception among Japanese consumers that imported food products are inferior in quality. This is an important point for exporters of food products to Japan. Japanese chauvinism may mean that imported products, even though they may be identical in quality to Japanese products, may be discounted in the marketplace.

In 1984 the U.S. Meat Export Federation (MEF) surveyed both meat shop owners and consumers with the aim of identifying the factors that influenced their decision to purchase beef (Hayes, 1990b). Forty-five percent of the meat shop owners and 40 percent of the consumers rated freshness as an important factor in their purchasing decision. For meat shop owners, 29 percent rated price as important, and nine percent rated cleanliness as important. Eighteen percent of consumers rated price as important, 11 percent rated safety of the product as important, and seven percent regarded cleanliness as important. The results clearly demonstrated the heavy weight that freshness has in the purchasing decision. When the consumer survey was divided among Japanese beef, chilled imported beef, and frozen imported beef, price increased in importance from 18 percent to 23 percent to 44 percent, respectively. This suggests that imported beef is not purchased for its physical characteristics but for its relative price.

Different Age Groups

An analysis of Japanese consumer preferences by age group suggested that younger people would like to increase their purchases of meat, particularly beef, rather than fish (Khan et al., 1990). Young men in their teens expressed the strongest preference for beef.

The analysis of various meat product preferences by men in different age groups suggested that while fish was most preferred by men in their fifties, preference for fish was inversely related to age (it was least preferred by men in their teens). Pork was the top choice of men in their teens and the second choice of men in their fifties. Chicken was preferred by just over 40 percent of men in their teens, and by approximately 20 percent of men in their twenties. Men in their thirties and fifties exhibited the lowest preferences for fish, where just over 10 percent of both groups preferred chicken. Beef was a popular meat with all age groups. Over 90 percent of men in their teens had a preference for beef. Approximately 80 percent of all other age groups (i.e., men in their twenties to fifties) preferred beef (Khan et al., 1990).

Khan et al. (1990) also reported data for meat preferences of Japanese women by age group. As with the men, women in their fifties most preferred fish. Pork was preferred slightly more by women in their teens than by women in their fifties, with only a small difference between pork preferences by the other age groups.

Women in their teens exhibited the strongest preference for chicken, with a slightly stronger preference for chicken over beef. Finally, women in their teens had the lowest preference for beef over the five age groups, with women in their thirties, forties, and fifties all expressing equally strong preferences for beef.

These results suggest that the older generations are more likely to eat fish. Of the four meat products listed, beef was the most popular for all male respondents, regardless of age, and for all women, apart from those in their teens who preferred chicken. Thus, beef is potentially a very popular meat in Japan. In the past, an expansion of beef consumption has been deterred, to a certain extent, by high beef prices and lack of availability. This survey did not, however, differentiate among different types of beef. Therefore, it may be that different age groups have stronger preferences for traditional marbled beef dishes whereas other age groups prefer more Western style dishes which use leaner beef.

CONSUMPTION AT HOME VERSUS
CONSUMPTION AWAY FROM HOME

The urbanization of the Japanese population has seen a boom in the restaurant trade. Yoshida and Klein (1990) report that the restaurant industry, on average, grew by 6 percent a year between 1975 and 1988. Several causes for this growth have been identified: the increase in disposable incomes; the increase in the number of young people with considerable purchasing power; the growth in nuclear families; the increase in the number of working women; more leisure time; the growth in daily dining out; and the diversification of meals (Yoshida and Klein, 1990; Khan et al., 1990).

Family restaurants and fast-food restaurants such as Denny's, McDonalds, Kentucky Fried Chicken, and Wendy's, have become more common. This is also partly a function of the internationalization of Japanese attitudes as more Japanese people travel abroad and the influence of media, particularly television, grows.

Sushi[3] is the most frequently chosen meal when eating out in Japan. Beef is the most popular meat consumed outside of the home, followed by pork. Poultry and processed meats are much less popular choices for out-of-home consumption (Khan et al., 1990).

Survey results reported in Hayes (1990b, p. 246) indicate that, in 1983, 39.5 percent of respondents selected beef as the meat for consumption out of the home. Twenty-seven percent consumed pork when eating outside of the home, and only 4.7 percent chose poultry and 2.5 percent chose processed meat. A full 10 percent of respondents would not select any meat product when eating outside of the home. In Western-style restaurants, beef tended to be consumed in the form of non-traditional steaks, roasts, and hamburger meat.

SUMMARY

The evolution of the Japanese diet has been influenced by a number of religious, historical, and cultural factors. Since the second World War, the Japanese diet has been influenced by Western dietary habits. This, together with an urbanizing and modernizing society, has led to a diversification in diet and in eating patterns. However, traditional Japanese preferences still heavily influence the demand for beef in Japan. Longworth (1983) comments that "Japanese people still maintain a strong preference for traditional Japanese cuisine" (p. 11) and that "Beef . . . cooked Japanese style, is still very much a special occasion food in Japan" (p. 12).

Chapter 3

The Dynamics of Japanese Population

Trends and changes in the dynamics of Japanese population will have a major effect on future Japanese beef imports. Changing demographic patterns will also have to be accommodated in the marketing strategies of those who wish to sell beef in Japan and will directly affect beef producers in exporting countries. It should be remembered that while per capita consumption of beef in Japan is relatively low compared to other developed countries, the Japanese population is also very large–more than 120 million. A one percent increase in the Japanese population translates into 11,280,000 more kgs of consumption.

Changing demographic patterns will also have ramifications for beef exporters. Younger consumers may be more receptive to trying new products and consuming higher quantities of Westernized products. Older consumers may be more likely to be set in their ways and want to limit their consumption to additional traditional Japanese dishes. Marketing strategies will have to be tailored to each group and different products made available to satisfy the divergent tastes. Westernized beef consumption will require imports of more lean product suitable for hamburger and Japanese adaptations of Western foods that use steaks and roasts as inputs. At the same time, traditional Japanese dishes will require the production of heavily marbled animals. While the table trade (home consumption) may require chilled beef, restaurants can often make do with frozen beef. In fact, they may prefer it if they have adequate freezer capacity. Since there are considerable regional differences in consumption, regional demographic patterns may also be of considerable importance to marketers as they plan their strategies.

GENERAL TRENDS IN JAPANESE
POPULATION GROWTH

The growth of the Japanese population during the post-war period is presented in Table 3.1. While population increased about 65 percent over the period, the most notable trend has been the declining *rate* of population increase. Between 1946 and 1955, the average annual rate of population increase was 1.9 percent; for the period 1980 to 1990 the rate had fallen to 0.7 percent. The major reason for this reduction has been the decline in the birth rate.

During the first quarter of the twentieth century, Japan could be characterized as having the high fertility levels and high mortality rates common to present-day developing countries. In 1925, women in Japan gave birth to an average of 5.1 children during their childbearing years, but this had fallen to less than 1.6 by the end of the 1980s (Martin, 1989). This decline in birth rates, however, has been partially offset by increasing life expectancy. In 1925, life expectancy at birth was about 45 years. By the end of the 1980s it had improved to 76 years for males and 82 years for females. As a result, a small net increase in Japanese population is projected for the immediate future.

Japanese population projections for the years 1995 to 2045 are presented in Table 3.2. In approximately 2015, population is projected to peak at 136 million. This represents only a 9 percent increase over the 1990 population. The additional 11.7 million people would, if they consumed beef at the current per capita rate, represent an increase of nearly 100 million kgs, or 275,000 extra cattle per year. Assuming that domestic Japanese beef production remains relatively unchanged, this represents a reasonable estimate of the minimum level of increased imports that exporting nations could expect.

Projections of total population increases, however, obscure significant demographic trends. One major trend is that the Japanese population is aging rapidly. The percent of Japanese over the age of 65 is expected to increase from its current 12 percent to 24 percent in 2020. This aging of the population has a number of ramifications for future beef consumption in Japan. In general, elderly people require less food than younger individuals. Greater amounts of

TABLE 3.1. Japanese population 1946-1990 (000).

Year	Population	Year	Population
1946	75,750	1969	102,536
1947	78,101	1970	103,720
1948	80,000	1971	105,145
1949	81,773	1972	107,595
1950	83,200	1973	109,104
1051	84,541	1974	110,573
1952	85,808	1975	111,940
1953	86,981	1976	113,094
1954	88,239	1977	114,165
1955	89,276	1978	115,190
1956	90,172	1979	116,155
1957	90,928	1980	117,060
1958	91,767	1981	117,902
1959	92,641	1982	118,728
1960	93,419	1983	119,536
1961	94,287	1984	120,305
1962	95,181	1985	121,049
1963	96,156	1986	121,672
1964	97,182	1987	122,264
1965	98,275	1988	122,610
1966	99,036	1989	123,120
1967	100,196	1990	123,540
1968	101,331		

Sources: Japan, *Statistics Bureau* (1988), *Japan Statistical Yearbook,* Management and Coordination Agency, Tokyo.

International Monetary Fund (1991), *International Financial Statistics,* Washington, DC.

TABLE 3.2. Japanese population projections 1995-2045 (000).

Year	Total	0-14 Years Old	%	15-64 Years	%	65 and Older	%
1995	127,565	22,387	(18)	87,168	(68)	18,009	(14)
2000	131,192	23,591	(18)	86,263	(66)	21,338	(16)
2005	134,347	25,164	(19)	84,888	(63)	24,195	(18)
2010	135,823	25,301	(19)	83,418	(61)	27,104	(20)
2015	135,938	23,876	(18)	81,419	(60)	30,643	(23)
2020	135,304	22,327	(17)	81,097	(60)	31,880	(24)
2025	134,642	22,075	(16)	81,102	(60)	31,465	(23)
2030	134,067	23,009	(17)	80,057	(60)	31,001	(23)
2035	133,133	23,914	(18)	78,278	(59)	30,941	(23)
2040	131,646	23,798	(18)	76,110	(58)	31,738	(24)
2045	130,017	22,809	(18)	75,824	(58)	31,384	(24)

Source: Japan, *Statistics Bureau* (1988), *Japan Statistical Yearbook*, Management and Coordination Agency, Tokyo.

protein are required for body development in young people, who are typically more physically active both in the workplace and during recreational activities. Further, at least in the intermediate run, these elderly people will be primarily very traditional or non-Westernized. Hence, even if there is a move toward greater Japanese acceptance of Western cuts of beef, it will not be by individuals moving into this age category. As a result, a significant portion of the population will still not have acquired a taste for large quantities of beef in their diet. What beef they will wish to consume will be of the traditional heavily marbled Japanese product, sliced paper-thin. The beef must be fresh, and retail portions must be small and processed almost to the table-ready stage, probably by a local butcher. Therefore, beef reaching consumers in this age category will have a great deal of value-added embedded in its cost, and the price will remain high, further limiting consumption. This is particularly true if the incomes of the elderly are reduced upon retirement.

Approximately two-thirds of Japanese people over the age of 65 live with their children in what is known as the traditional "stem" family structure. When the eldest son marries he brings his wife into his parents' home and the two generations live together until the parents' death. The eldest son inherits all the family assets. Daughters and younger sons are left to establish their own households. While this traditional stem structure has weakened recently due to the need for mobility among young people seeking better employment opportunities, this is largely a delay rather than a fundamental change. The result has been that the state has provided fewer services for the elderly than in most Western countries. As children have been expected to care for their parents, this has meant little urgency to provide pensions whereby the elderly can independently provide adequately for their own well-being upon retirement. This is further aggravated because the relatively short lifespan of the past did not require heavily funded pension plans. Japanese pensions tend not to be for the remaining life of the retiree but rather for some fixed period, typically ten years. Often it is paid out in a lump sum at the time of retirement. While the amount will vary depending upon years of service, sex, and education, it typically amounts to about four years of the individual's last basic salary. The government has been working to expand the role of public pension plans but, as yet, the major responsibility remains with the children.

How will the decline in income and independence for an increasingly aging population be reflected in the future demand for beef? While most elderly Japanese eventually live with their children, there is an increasing period of independent retirement. As is well known, the Japanese have very high savings rates, but the inadequacy of existing pension plans still suggests that there will be a considerable reduction of income upon retirement. This is borne out by the high rate of part-time employment among the elderly. In 1987, 36 percent of males and 15 percent of females aged 65 and older were working (Martin, 1989). As a result, it would seem that consumption of high-priced beef would be unlikely to increase upon retirement and, in fact, would most likely decrease. This is particularly true given that the type of beef this age group is likely to consume will always be expensive. This is simply because of the high feed costs associated with producing a high degree of marbling and the high level of value-added required to prepare the product for the home consumption market. Hence, it would seem that beef will remain a "special occasion" dish for a significant portion of the elderly market.

When the move to the children's home eventually takes place, the effects on beef consumption are likely to be more pronounced. First, there is likely to be some cross-subsidization within the household as the children support the parents' expenditures. One of the simplest and most "face-saving" methods of doing this is to assume the lion's share of the household's common food bill. As a result, the Japanese housewife will be faced with a considerable strain on her food budget. This would suggest a substitution away from high-priced beef. Further, the presence of the elderly may arrest any trend toward consumption of more Westernized beef dishes. As the elderly are less likely to have acquired a taste for Westernized beef, the daughter-in-law is less likely to purchase it for meals consumed with the parents. If the older generation female assumes the responsibility for shopping, she may not be familiar with Western beef and therefore not purchase it. All of these effects would seem to suggest a moderating influence on beef consumption as the population progressively ages.

Even if there is a significant increase in the proportion of the elderly who do not live with their children, the effects on beef consumption will still tend to be moderated. The likely alternative to living with one's children will be some form of institutionalization, as has occurred elsewhere in the developed world. The degree of institutionalization tends to increase with age, and along with progressive institutionalization comes institutional food. Institutional food for the elderly tends to be characterized by blandness, traditional dishes, and constant attempts to minimize costs. While low-quality frozen beef may be relatively inexpensive in the wake of liberalization, it will not fit the other two criteria. Certainly, institutions are not likely to purchase the high-quality product that appears to have the greatest profit potential for exporters.

One consequence of the rapidly aging Japanese population is that growth in beef consumption will likely be somewhat constrained (1) because elderly people eat less beef and (2) because working people will be spending a greater proportion of their income supporting the elderly. As a result, demographically, major increases in beef consumption can be expected to come only from the young who are not yet caring for their parents.

Labor market participation among the young is increasing, suggesting more affluence and a better acquaintance with the ways of the West. A commonality of culture—music, movies, dress, etc.—will also spill over to food to some extent. Just as North American children are much more comfortable with pizza, tacos, and sushi than their parents, so too will Japanese youth be more comfortable with hamburgers, steaks, and prime rib. Many of them will have travelled abroad. This is not to suggest that they will turn their backs on Japanese cuisine, but that they will be much more cosmopolitan and ethnically diverse in their food consumption. The major question is not whether their tastes will become more globalized but, rather, what form they will take.

The hamburger trade requires an entirely different product than does the steak trade. Hamburger and hamburger-type foods—chicken-fried steak in a bun, roast beef sandwich, beef dip—are not eaten with chopsticks. They are eaten with the hands, which is a far easier transition to make than to the knives and forks required by the steak trade. Further, new foods tend to be learned by young people in fast-food restaurants. North American youth learn about "knife and fork" cuisine in the home. It may be relatively easy to translate the fast-food experience to the Japanese home, but the large cuts of beef will be much more difficult to move into the home environment. Further, the Japanese do not like the thought of leftovers, so that roasts, long the multiple meal mainstay of Western beef consumption, have limited usefulness in Japanese household menus. The net result is likely to be a significant increase in the hamburger trade, some modest increase in the steak trade, and little or no market for large cuts.

Of course, the relatively high incomes of younger Japanese–although somewhat moderated by the Japanese system of basing pay on years of service rather than skill or ability–may allow them to consume larger quantities of well-marbled beef in traditional dishes. Over the next 15 years some 23 million Japanese will have high discretionary incomes, and they would seem to be the group to which beef promotions should be targeted. If their tastes can be altered, then they are likely to retain those tastes throughout their lives.

REGIONAL DIFFERENCES IN DEMOGRAPHIC TRENDS

Population density varies tremendously by prefectures, ranging from 72 persons per square kilometer in the most northerly prefecture of Hokkaido to 5,571 in Tokyo. The average population density is 325 persons per square kilometer. Population density has considerable implications for marketing and promotions as it affects everything from the efficiency of marketing expenditures to traffic in supermarket aisles. It also affects the types of retail outlets that can prosper. Small independent butchers are more likely to survive in heavily populated areas. Of course, this has implications for beef distribution. Butchers require small quantities of beef that are in a form suitable for transport by small motorcycles. Further, independent butchers may not have the refrigeration necessary to handle frozen product.

High population density areas have higher rents that are eventually factored into the markup on the product, thereby increasing cost to the consumer. Further, if rents and real-estate prices rise, more of a consumer's income would be required for housing costs and mortgages, reducing money available for luxury food items such as heavily marbled beef or the large cuts associated with North-American-style beef.

The fastest growing prefectures in the first half of the 1980s were Nara (6.53 percent) and Shiga (6.51 percent) in central Honshu. Tokyo had only a 0.15 percent growth rate, reflecting its already crowded nature. The prefectures surrounding Tokyo, however, all had healthy population growth rates. Only one prefecture, Akita in northern Honshu, had a decrease in population. In general, the greatest growth tends to be exhibited in central Japan, with progressively less growth at the extremities. This is the result of internal migration rather than birth rates since urban areas have lower birth rates than rural areas.

URBAN VERSUS RURAL POPULATION

As with most developed countries, the Japanese population is rapidly urbanizing. Seventy percent of the population lives in urban areas, and 43 percent of the population is concentrated in the cities and surrounding areas of Tokyo, Osaka, and Nagoya. Agriculture employs only about ten percent of the labor force. The vast majority of these people hold part- or full-time jobs off the farm.

Urban populations tend to have much more cosmopolitan eating habits. In part, this is a result of accessibility to foreign food via the restaurant trade. Urban consumers also tend to have higher incomes, which allow more latitude in food experimentation. They also tend to be more susceptible to fads and, hence, can be more easily convinced to experiment. The urban dweller is also less likely to have a traditional home. While working wives are not common in Japan, particularly during the child-rearing years, they are on the increase. Families with two working members tend to have higher incomes but have more time constraints. As a result, they opt more for convenience foods with higher degrees of value added but which require less preparation time. Japan, as with much of the developed world, has been overtaken by the microwave revolution.

Convenience foods rely on a high degree of processing. As a

result, the quality of beef used in convenience foods becomes less important. Marinating, pre-cooking, reconstituting, flavoring, etc., along with the heavy use of ground product, reduces the need for a tender or inherently tasty product. Freezing beef adds a dimension of storage flexibility that is not possible when chilled beef is required as an input. Hence, the processed market can cope quite well with manufacturing quality beef in frozen form. Further, in many cases the base meat for processed product is not important. Substitution among pork, chicken, lamb, and beef is often possible. This makes the level of beef sales much more dependent on price, particularly price relative to competing meats. Hence, to some extent the urbanization of the Japanese population is likely to encourage the importation of low-quality beef. As a result, the success of beef sales will be much more dependent on its price competitiveness than on its quality. On the other hand, as urban dwellers tend to be more sophisticated and can more easily shop competitively, they will likely demand the highest quality at a competitive price.

SUMMARY

Demographic trends in Japan are likely to have a significant influence on the quality as well as the quantity of beef consumed in Japan. While the Japanese population will continue to grow in the near term, the rate of growth is expected to continue to decline. The population itself is projected to eventually decline by the end of the first quarter of the twenty-first century. The Japanese population is growing in a particular way. It is aging. This will have direct effects on beef consumption because older people tend to eat less and have more conservative eating habits. This suggests that the type of beef consumed will be of the more traditional, heavily marbled type.

The rapidly aging population will also require ever-increasing resources from the working population for its maintenance. Further, these older people will be living longer with their children. This means less discretionary family income and a possible redirection of beef consumption toward more traditional dishes. As a result, only the young working population will have large discretionary incomes with which to experiment. Unfortunately, this is the portion of the labor force that is growing least quickly.

Chapter 4

Income and Beef Consumption

One of the most long-standing food consumption patterns is that as an individual's income increases, the amount spent on food consumption also increases. This has been found to hold across societies and over time. It should be noted, however, that the quantity of food consumed does not necessarily increase as income increases. While it may be true that quantity increases with income when incomes are very low, at higher levels income increases begin to have an effect on the composition of the food basket rather than the total quantity of food eaten.

There are five major ways in which income increases can affect the composition of the food basket. First, there is a substitution of higher-priced food groups for low-priced staples. In most cases this means that the nutrient quality of the food consumed increases. Specifically, it means the movement away from carbohydrates and toward the greater consumption of foods with higher levels of meat protein. Second, there is a broadening of the food basket to include a wider variety of foods. Third, consumption tends to progress through phases from dried to canned to frozen to fresh. Fourth, further processed foods or convenience foods replace foods that require a great deal of preparation or cooking time in the home. Finally, food consumption in restaurants partially replaces food consumed in the home.

The measure of the response of consumption to an income increase is known as income elasticity. It is defined as the percentage change in quantity consumed divided by the percentage change in income. Most agricultural commodities have a value between 0 and 1, which means that while consumption increases as income increases, it increases at a smaller rate than the income increase.

THE EVOLUTION OF THE JAPANESE DIET

While any diet is partially determined by culture and–in a pre-trading economy–the available food, it is the removal of income constraints that allows for dietary evolution. The rapid increase in Japanese incomes since World War II has led to a rapidly evolving diet. As the increase in income has accelerated through the late 1970s and 1980s, so too have the changes to the Japanese diet.

Before World War II, approximately 60 percent of caloric intake in the Japanese diet came from rice. Rice can be considered a premium dietary staple since it contains most of the nutrients required for a balanced diet. Rice, combined with miso soup, and augmented at times with vegetables and small quantities of dried fish with salt, was the typical daily diet.

The evolution of the Japanese diet over the period 1960-1990 is illustrated in Table 4.1. Within food groups the substitutions for the source of both calories and protein follow a pattern common to societies whose incomes rise. In terms of energy sources there has been a general decline in the role of cereals, from almost two-thirds of caloric intake to not much more than one-third. This reduction in cereal intake was largely replaced by an expanded consumption of dairy products and meat. Hidden in the other category is also a large increase in the consumption of fats, largely from cooking oils. In 30 years, meat (excluding fish) and dairy product consumption rose from 4 percent of caloric intake to 15 percent, suggesting a radical shift in diet.

An examination of protein sources is even more indicative of the effect of income on diet. The general movement from vegetable to animal sources of protein reflects the relative costs of potential components of the food basket. Due to the relatively inefficient conversion of vegetable protein that is a fact of life in animal production, meat-based protein sources tend to be relatively expensive. In 1960, protein from cereals and beans comprised 53 percent of the total protein intake, while that from meat and dairy sources constituted only 8 percent. By 1990, meat and dairy products had contributed over 30 percent of the total protein intake, whereas cereal and beans had declined to 35 percent. Fish and seafood remained fairly constant at approximately 22 percent of protein intake.

TABLE 4.1. Changes in the composition of the Japanese diet 1960-1990.

Food Group	Sources of Calories (Percentage)		Food Group	Sources of Protein (Percentage)	
	1960	1990		1960	1990
Cereals	63	38.7	Cereals	41	26.2
Fruit and Vegetables	9	5.2	Fish	22	21.4
Starchy Staples	6	7.7	Beans	12	8.8
Fish	4	5.1	Eggs and Dairy	6	14.1
Eggs and Dairy	3	8.2	Meat	2	16.2
Meats	1	6.8	Other	17	13.3
Other	14	28.3			

Source: Japan MAFF (1992).

Percentages, however, hide another salient feature of the normal evolution of diets as incomes increase. Table 4.2 provides a breakdown of protein consumption (in grams) by source per day for Canada and Japan. Canada can be considered a developed economy that has had a high standard of living for a considerable period. As a result, its dietary pattern can be considered mature. This contrasts with Japan whose rapid income increases over the last three decades have moved it into the category of developed economy. Of course, per capita income in Japan has recently become higher than in Canada.

As incomes rise, so does the general level of protein consumption. Between 1960 and 1985, Japanese protein consumption rose from 76 percent of Canadian levels to 85 percent. Protein from animal sources increased from 30 percent of the Canadian level of

TABLE 4.2. Protein supply per capita per day–Japan and Canada.

| Food Group | Grams per Day 1960 and 1985 | | | |
| | 1960 | | 1985 | |
	Japan	Canada	Japan	Canada
Meats	2.8	29.3	12.4	33.9
Eggs	2.2	4.6	5.0	3.6
Milk and Milk Products	1.8	24.6	5.3	22.7
Subtotal Animal Products	5.6	58.5	22.7	65.5
Fish and Shellfish	14.6	2.3	18.5	2.4
Subtotal Animal Protein	20.2	60.7	41.3	67.9
Cereals	28.8	21.5	23.8	22.6
Beans	8.0	1.2	7.4	1.4
Others	11.5	7.4	11.6	8.9
Subtotal Vegetable Protein	48.3	30.1	42.8	32.9
TOTAL	68.8	90.8	84.1	100.7

Sources: Japan MAFF (1992).
          Canada, Statistics Canada, Food Consumption Statistics, Ottawa.

intake to 60 percent. The most dramatic increase, however, was the one in protein from meat, from six percent of the Canadian intake in 1960 to 37 percent in 1985. In general, as Japanese incomes have risen there has been a convergence in the two diets.

Much is made of the fact that the Japanese consume less food than North Americans and that being overweight carries a considerable stigma. Of course, over-consumption of animal fat is one of the major reasons for many North Americans being overweight. In general, if one allows for the dislike of obesity in Japan and the relative prices of the various protein sources, it would appear that the Japanese diet now approximates the diets in other developed countries.

CHANGES IN INCOME AND BEEF CONSUMPTION

While meat consumption has been rising, Japanese beef consumption remains far below that of other countries with similar levels of income. In general, there is a high correlation between the level of per capita income and the consumption of beef (see Table 4.3). One obvious exception is Japan.

While beef consumption remains low in absolute terms, what is really important is the responsiveness of beef consumption to income increases. While estimates of the income elasticity of beef purchases in Japan vary considerably, there is general agreement that they are large relative to those in most developed countries. Values range from 0.9 to 1.89 (Simpson et al., 1985), which means that a 10 percent increase in real incomes would lead to an increase in quantity of beef demanded of between 9 and 18.9 percent. Most of the estimates are greater than one. This puts beef in Japan in the category of a "luxury good," where consumption increases at a faster pace than income. This is typical of high-priced goods with low levels of consumption. As the level of consumption rises, one would expect the income elasticity to fall as satiation is approached. However, given the relatively low level of current consumption, satiation is likely to be a long way off. Thus, the important question becomes what are the prospects for the continued growth in Japanese income?

PROSPECTS FOR CONTINUED GROWTH IN JAPANESE INCOMES

The so-called Japanese economic miracle is one of consistent economic growth. Table 4.4 shows the rates of growth in the Japa-

TABLE 4.3. Meat and beef consumption per capita and per capita income in various countries, 1989.

Country	Per Capita (kg/yr) Beef[1] Consumption	Per Capita (kg/yr) Meat[2] Consumption	GDP Per Capita[3] in U.S. $**
U.S.	45.9*	120.8*	20,635
Japan	8.1	43.4*	23,329
Canada	39.3	102.1	20,740
Australia	42.7*	107.5*	16,863
New Zealand	37.4	98.6	12,970
Denmark	19.1	105.6	19,867
France	30.4	109.6	17,143
W. Germany	22.7	100.0	19,084
Greece	22.1	82.4	5,400
Ireland	19.1	88.8	9,857
Italy	26.6*	88.8	15,118
U.K.	19.1	73.8	14,589
Turkey	7.9*	22.9*	1,433
Portugal	14.0	70.5	4,398

Sources: 1. OECD (1991).
2. OECD (1992a).
3. OECD (1992b).

*Estimate by the OECD Secretariat.
**1991 prices and exchange rates.

nese economy. While the double-digit growth rates of the 1960s may not be attainable in the 1990s, the Japanese economy still exhibits a faster and more sustained rate of growth than do the economies of other countries. Even in the worldwide economic slowdown of the early 1990s, the rate of growth in the Japanese economy outpaces that of most other developed countries. If the average rate of income growth in the 1980s (approximately 4.5 percent) can be sustained in the 1990s, then one could expect an annual growth rate in beef consumption of approximately 4 percent, using the most conservative estimates of income elasticity. There is no reason why such growth rates cannot be expected since the

TABLE 4.4. Growth rates in real GDP—selected OECD countries (percentage changes from previous year).

	Average					Annual		
	1960-73	1974-79	1980-82	1983-87	1988-89	1988	1989	1990
Japan	9.9	3.6	3.7	4.1	5.3	5.7	4.9	4.7
United States	3.8	2.6	–0.3	4.0	3.7	4.4	3.0	2.3
Germany	4.7	2.4	0.2	2.2	3.8	3.6	4.0	3.9
OECD Europe	4.6	2.9	0.9	2.5	3.6	3.8	3.5	2.9
Total OECD	5.0	2.8	1.0	3.4	4.0	4.4	3.6	2.9

Source: OECD (1990).

Japanese economy is better balanced than in the past. This is the result of a major restructuring combined with strong and continuing investment in research and development.

SUMMARY

The available evidence suggests that beef consumption in Japan is very responsive to increases in consumer income. Hence, if Japanese incomes can be expected to continue to increase, the prospects for increased beef consumption and imports would appear to be excellent. As yet, there is no reason to predict a significant decline in the rate of Japanese economic growth.

Chapter 5

Government Regulations
for Livestock and Meat

One of the most important aspects of marketing agricultural commodities is ensuring compliance with government food regulations. Failure to comply with regulations can lead to rejection of product, loss of purchaser confidence, and possible litigation from consumers that have been sold products in contravention of the regulations. The regulations affecting food tend to differ considerably from country to country. As a result, exporters may find that procedures and practices acceptable for products to be sold in the domestic market will not be acceptable in the importing country. At the least, this means additional costs related to retooling production lines, training of production and inspection staff, and redesign of packaging materials. Sometimes it is not possible to simultaneously satisfy the regulations of two countries in one plant and a separate facility may be required for the export market. On rare occasions it may not be possible to satisfy the domestic and importing country's regulations at the same time, so export sales are foregone entirely. Hence, it is extremely important, before any marketing attempt is undertaken, that those charged with opening new markets become acquainted with the importing country's regulations. They then must meet with both those responsible for producing the product and domestic government officials who are responsible for inspection of exports. Of course, officials from the importer's inspection branch also need to be met and a good working relationship established.

Familiarity with regulations is particularly important for live animals and extremely perishable products, such as meat, since regulations tend to be very stringent due to potential hazards to both consumers' health and the importers' livestock industry. Regulations can generally be divided into three broad categories: (1) protection of health; (2) sanitary regulations; and (3) consumer protection. Each of these are discussed in the following sections.

PROTECTION OF HEALTH

The regulations that apply to domestic animals are often reflected in the quarantine regulations of an importing country, as the intent of quarantine regulations is to ensure that imported animals meet the same health standards as do domestic animals.

Animal Quarantine in Japan

Japan is one of the few countries in the world where all serious animal diseases have been eradicated. Highly infectious animal diseases such as foot-and-mouth disease, rinderpest, African swine fever, etc., still persist in many parts of the world. An outbreak of any one of these diseases would have considerable economic costs associated with it. In many cases the diseases are untreatable–individual animals or whole herds have to be destroyed if infections occur. Often it takes years to ensure that the disease has been eradicated. As a result, a country whose animals have the disease may lose export markets for everything from meat to breeding animals to semen. Characteristics of the major serious infectious diseases affecting animals are shown in Table 5.1.

In order to prevent infection or re-infection through imports from areas where the diseases are active, the Japanese government has imposed restrictions on the importation of live animals and their products. Import and quarantine regulations are administered under the Domestic Animal Infectious Diseases Control Law. Japanese regulations are similar to those of other countries that have eradicated major livestock diseases.

Animals and animal products subject to quarantine are termed "designated quarantine articles" and are subject to both import and export

TABLE 5.1. Characteristics of major exotic and serious animal diseases.			
Name of Disease	Causative Agent	Symptoms	Distribution
Foot-and-mouth disease (FMD)	FMD virus (7-types and more than 60 sub-types of viruses already discovered)	The disease (FMD) is an acute, highly communicable disease affecting almost exclusively cloven-hoofed animals, domesticated and wild. It is characterized by high temperature and the formation of vesicles and erosion in the mucosa of the mouth and the skin above the claws of the feet. There are reductions or cessations of milk flows, abortions, and emaciation with a significant reduction in the value of the animal.	Most of the countries in the world except Japan, North America, Central America, Australia, New Zealand and some countries in Europe.
Rinderpest	Rinderpest virus	Rinderpest (cattle plague) is an acute, highly communicable disease, primarily of cattle and buffaloes, secondarily of sheep, goats, and wild ruminants. The disease is characterized by an abrupt rise in temperature, erosions on the gums, diarrhea, and death in 7 to 12 days.	Enzootic in Africa and Asia. Japan is free of the disease.
African swine fever (ASF)	ASF virus	African swine fever is a highly contagious, often acute disease of swine and wild hogs, characterized by fever, hemorrhages of the internal organs and death in 5 to 15 days. Mortality frequently approaches 100 percent.	Most of the countries in Africa and Southern Europe – Spain, Portugal, Italy, etc. Japan is free of the disease.

quarantine to ensure that the causative agents of infectious animal diseases are not disseminated. Such animals must be "passed" by an animal quarantine officer of the Animal Quarantine Service. Table 5.2 provides a list of the major species that are subject to quarantine regulations; Table 5.3 provides a list of animal products subject to the regulations.

Import Prohibition

Import prohibitions are used to prevent the introduction of infectious animal diseases into Japan. Certain products from areas designated as "import-prohibited" can never be imported into Japan, even if accompanied by an inspection certificate.

As shown in Table 5.4, all countries are classified into four categories, as determined by the state of animal health in the country. Anyone may import cloven-hoofed animals, their semen, embryos, and meat viscera, and ham, sausage, and bacon manufactured from the meat or viscera from countries in the group designated as 0. Cloven-hoofed animals, their semen and embryos, as well as ham, sausage, and bacon manufactured from their meat or viscera are allowed to be imported from those countries in group 1. However, imports of meat and viscera from these countries are not allowed. Permission may be granted to import heat-processed canned or manufactured meat items from countries in group 1.

Cloven-hoofed animals, their semen or embryos, as well as heat-processed meats and other products produced under stipulated conditions may be imported from group 2 countries. However, meat, viscera, ham, sausage, and bacon cannot be imported from these countries.

Only heat-processed meats and meat products can be imported from countries that are designated as group 3.

Inspection Procedures for Animals and Animal Products

Upon arrival at an airport or seaport through which imports are permitted, animals are subject to isolation at a quarantine facility

TABLE 5.2. Designated animals for import-export quarantine in Japan.		
Classification in Law	Examples of Designated Animals for Quarantine	Examples of Animals Not Requiring Quarantine
Cloven-hoofed animals	Cattle, buffaloes, zebu, goats, sheep, hogs, wild hogs, hippopota-muses, giraffes, deer, camels, moose, llamas, alpaca, etc.	Elephants, bears, lions, pandas, monkeys, foxes, tigers, lizards, crocodiles, alligators, snakes, turtles, fishes, etc.
Horses	All varieties of horses, asses, mules, zebras, etc.	Rhinoceroses, etc.
Chickens, quail, turkeys, ducks, and geese	All varieties of chickens, quail, turkeys, ducks, and geese	Pheasants, cranes, peacocks, hawks, eagles, crows, etc.
Rabbits	All varieties of rabbits	Squirrels, guinea-pigs, mice, rats, etc.
Honeybees	Honeybees	Insects other than honeybees

and to an inspection conducted by an animal quarantine officer. These must be completed prior to customs clearance.

When an animal is found to be infected or is suspected of having an infectious animal disease, the animal may be isolated for further observation, then slaughtered and possibly incinerated, if required.

An inspection certificate issued by the government authorities of an exporting country must accompany a designated quarantine article when imported into Japan. The certificate is required regardless of the amount involved and the purposes for which it will be used. The certificate must state that, as a result of the inspection by the appropriate authorities in the exporting country, it is believed that the designated quarantine articles will not disseminate the causative agents of infectious animal diseases. The inspections undertaken by

TABLE 5.3. Designated animal products for import-export quarantine in Japan.

Classification in Law		Examples of Designated Animal Products	Examples Not Requiring Quarantine
Chickens, ducks, turkeys, quail, and geese	Eggs	Breeding eggs, table eggs, frozen eggs, etc.	Boiled eggs and eggs of ostriches, canaries, etc.
Cloven-hoofed animals, horses, chickens, ducks, turkeys, quail, geese, rabbits, and honeybees	Bones	Raw, dried, steamed or crushed bones, bone powder, etc.	Bone decorations finished completely, ivory, etc.
	Meat	Fresh, frozen, chilled, salted, boiled, steamed or dried meat, meat powder, etc.	Meat derived from kangaroos or whales
	Fat	Fresh, frozen, or salted fat	Finished lard, etc.
	Blood	Fresh or dried blood, sera, etc.	
	Hides and skins	Raw, salted, dried or cured hides and skins	Leather
	Hair	Raw, washed or not completely finished hair, mohair, cashmere, etc.	Carbonized wool, knitting wool, etc.
	Feather	Raw or not completely finished feathers	Completely finished feathers
	Horns	Horns	Completely finished horns

Classification in Law		Examples of Designated Animal Products	Examples Not Requiring Quarantine
	Hooves	Hooves and crushed or powdered hooves	Completely finished hooves
Cloven-hoofed animals, horses, chickens, ducks, turkeys, quail, geese, rabbits, and honeybees (cont'd)	Tendons	Fresh, frozen, dried, or boiled tendon	
	Viscera (internal organs)	Fresh, frozen, chilled, dried or boiled viscera, casings, etc.	
	Raw milk	Raw milk	Butter and cheese
	Semen	Fresh or frozen semen	
	Embryos	Fresh or frozen embryos	
	Blood powder	Blood powder or serum powder	
	Ham	Ham	
	Sausages	Sausages	Fish Sausages
	Bacon	Bacon	

the appropriate authorities in the exporting countries must be agreed to by the Japanese animal health authorities.

SANITARY REGULATIONS

The major laws relating to Japanese sanitary regulations for food are the Abattoir Law and the Food Sanitation Law. The intent of the

Abattoir Law is twofold: (1) to promote the proper handling of animals; and (2) to contribute to the improvement of public health. The animals covered are cattle, pigs, sheep, goats, and horses. The Food Sanitation Law regulates the safety of food and food additives. It was enacted on January 1, 1948, and is intended to decrease the risk of any hazard that can arise from the poor handling of food and to promote public sanitation.

The law forbids the harvest, production, selling, importing, processing, cooking, and storing of either food or food additives which are (1) rotten (2) harmful (3) contaminated with a virus, or (4) unclean. The sale of the meat, milk, blood, guts, or bones of diseased animals is also illegal. Food additive use is strictly controlled and all food additives must be certified before they can be used.

The Minister of Health and Welfare can make regulations regarding the methods of cooking, utilization, processing, and production of food and food additives. If food does not satisfy these regulations, it cannot be imported or sold. Packaging is regulated and any materials that present a health risk are banned. Food must be labelled with an indication that it has passed inspection.

Government food sanitation inspectors are located in prefectures and cities. Private firms that process food must use a food sanitation inspector who has medical training. The increasing quantities of food imports has led to an expansion of the food inspection service and food sanitation inspectors are now stationed in quarantine offices at all ports of entry.

Grades for food, manufacturing standards, and regulations regarding food and food additives have been established to protect public health. It is illegal to sell food and food additives that violate these standards. Chemical compounds can also be designated as food additives. The sale of undesignated additives, and foods containing any undesignated additives, is banned.

Standards have been established for restaurants and other businesses whose handling of food could have serious ramifications for public health. In the name of public health, the government can order managers to cease operation if the standards are not strictly followed. In order to protect against food poisoning and to prevent the sale of illegal foods, food sanitation inspectors visit establishments on a regular basis.

TABLE 5.4. Classification of import-prohibited areas.

Goods		Classification of Areas			
		0	1	2	3
Of cloven-hoofed animals	Live animals	Republic of Korea, Taiwan, Finland, Sweden, Norway, Denmark, Northern Ireland, Ireland, Iceland, Madagascar, Canada, U.S., Mexico, Guatemala, Nicaragua, Costa Rica, Northern Mariana Islands, New Zealand, Vanuatu, New Caledonia, and Australia	Singapore, Poland, Hungary, Netherlands, Belgium, France, Austria, Romania, Yugoslavia, Switzerland and United Kingdom (restricted to Great Britain only)	China and Federal Republic of Germany	Areas other than those mentioned under classification areas 0, 1, and 2
		Note 1	Note 1	Note 1	Note 2
	Semen and embryos	Note 1	Note 1	Note 1	Note 2

TABLE 5.4. Classification of import-prohibited areas (cont'd).

		Classification of Areas			
	Ham, sausage, and bacon	Note 1	Note 1	Note 3	Note 4
	Meat and viscera	Note 1	Note 2	Note 3	Note 4
Causative agents of animal infectious diseases	Note 2	Note 2	Note 2	Note 2	Note 2

The classification of areas in this table is amended in accordance with a change of animal health situation in exporting countries.

Note 1: Articles concerned may be imported if accompanied by an inspection certificate issued by the government authorities of exporting country.

Note 2: Importation is prohibited, except when permitted by the Minister of Agriculture, Forestry and Fisheries of Japan for special reasons, e.g., for experimental and research purposes.

Note 3: Importation may be permitted on condition that the products were processed, in accordance with the heat-processing standard stipulated by the Minister of Agriculture, Forestry and Fisheries of Japan, at an establishment designated by the government authorities of an exporting country as one conforming to the standard of heat-processing stipulated by the Minister of Agriculture, Forestry and Fisheries of Japan.

Note 4: Importation may be permitted on condition that the products were processed in accordance with the heat-processing standard stipulated by the Minister of Agriculture, Forestry and Fisheries of Japan, at an establishment designated by the Minister of Agriculture, Forestry and Fisheries of Japan as one conforming to the standard of heat-processing stipulated by the said Minister.

The system of sanitary regulations in place in Japan is similar to that in most developed countries. Japanese consumers are very concerned with food safety, and standards are strictly enforced.

CONSUMER PROTECTION

The final area of government regulation is in the area of consumer protection. The date of manufacture must appear on all processed food products. Labelling standards for food additives have been established. The government is intending to establish grading standards for 24 food groups, including meat. In June 1988, the Food Sanitation Research Committee recommended that it would be desirable to have food additives labelled. In 1988 and 1989, 347 chemical or synthetic food additives and 1,051 non-chemical food additives were required to be reported on packages if used in the manufacture of processed food.

Consumers are assured of the quality of their food by the extensive use of the Japanese Agricultural Standard (JAS) mark. The JAS mark is widely used as a criterion by consumers when making purchases, as both a standard of reliability in commodity transactions and a guideline for making quality improvements. The JAS mark can be applied to both domestically produced and imported foodstuffs. Those wishing to export further processed products containing beef to Japan may seek JAS grading and approval.

According to the JAS Standards System, the Minister of Agriculture, Forestry and Fisheries sets standards for individual commodities. Third-party inspection of products manufactured to JAS standards is conducted. The symbol can be used only for products that pass the inspection.

The government guarantees that all products that use the JAS mark meet or exceed JAS standards. This provides consumers with a considerable degree of security at the time of purchase. However, use of the JAS mark by manufacturers is voluntary, and distribution of products not displaying the JAS mark is not regulated.

The Minister of Agriculture, Forestry and Fisheries may, in order to achieve the goals of the JAS system, designate categories of products within the JAS system and establish standards for these products. To prevent JAS standards from becoming trade barriers, notification is given according to the procedures stipulated in the GATT Standards Code.

Food products imported into Japan are inspected according to the Food Sanitation Law and the Customs Law. If the product passes these inspections and has been JAS graded, then it is deemed to meet all regulations required by Japanese law. The JAS standards are developed within the "International Food Standards" (Codex Standards) of the FAO/WHO Codex Alimentarius for foods and drinks.

The inspection and grading system is well accepted by consumers. Over 90 percent of consumers indicate an awareness of the system, and almost 70 percent of consumers feel that it is an indication of high quality.

SUMMARY

Since Japan has high food safety standards, it is essential that those who wish to export their products to Japan become familiar with the details of the regulations and establish working relationships with appropriate regulatory officials. It is fundamentally important that beef leaving the exporting country conform to the health and safety regulations of Japan. Failure to do so can lead to financial losses in the short run and loss of markets in the long run through an erosion of the exporter's reputation.

Chapter 6

The Substitutes

While changes in the prices of many food and non-food items can be expected to affect Japanese beef consumption, a few close substitutes are likely to be of particular importance. The main substitutes for beef are those products which compete for that portion of the consumer's budget allocated for foods with high protein content. In other words, it is not likely that changes in the price of rice or noodles will have much effect on consumption of beef. Rather, Japanese consumers' purchases of beef will be influenced to a considerable degree by the prices of pork, chicken, seafood, and soybean-based products.

Other meats such as lamb and turkey have never been important in the Japanese diet. In the period when beef imports were very restricted, lamb and mutton imports grew somewhat as a source of inputs for processed meats. However, as beef import liberalization progressed, sheep-based products have gradually been replaced.

Turkey consumption, as in much of the world, has been on the increase. This is primarily because it can be produced and processed less expensively than chicken (Clarke, 1989). The price of turkey in Japan fell approximately 50 percent in the late 1980s and undersells chicken by as much as 30 yen per 100 grams. As in North America, much of the new success of turkey has arisen in the area of processed spiced meats, in products where the input meat does not much matter to the final product and, hence, substitution can take place within the preparations. Frozen turkey sausages and turkey burgers have made their appearance in Japan. In the upscale market, turkey drumsticks and ground meat are sold as specialty items. However, with imports reaching only 2,000 tonnes, turkey remains a very minor competitor for consumers' expenditures on high-protein foods. Consumer attention remains focused on the four major meat sources and soybeans. It is these products on which this chapter focuses.

PRODUCTS BASED ON THE SOYBEAN

Beans, and in particular processed soybean product, represent the only major high-protein source from vegetables. Beans traditionally represented the major protein source due to religious, particularly Buddhist, prohibitions against the killing of animals.

Soybeans are consumed in three major forms in Japan. The first is in fermented form as the flavoring agent for miso soup. While declining in popularity, miso soup remains one of the staples of the Japanese diet and is consumed as an integral part of most meals. Miso soup provides the major protein base for the traditional Japanese diet.

The second major form in which the Japanese eat soybeans is tofu (bean curd). The tofu industry is characterized by small-scale production and by being one of the most traditional industries. As a result, both the skills and resources to market tofu extensively are lacking. Partly for this reason and partly due to changing tastes, tofu has been losing market share to meat. Finally, a significant proportion of domestically produced soybeans are cut before they mature and are consumed as either a green vegetable or as an appetizer (Longworth, 1983).

Of course, soybeans are a dual purpose commodity. They are used not only for human food but also as a major protein source for livestock feed. The Japanese government has permitted feed grains, including soybeans, to be imported cheaply to support domestic meat and dairy production. Instead of supporting domestic producers through import controls, soybean farmers have been supported through deficiency payment schemes that cover the difference between the required domestic price and the price that farmers receive when they sell their soybeans in the market. In effect, the world or import price has been the effective price that soybean users pay in Japan.

Since livestock feed prices, including soybeans, tend to fluctuate considerably in the world market, the Japanese government has operated a storage scheme whereby buffer stocks are maintained. These stocks are administered by the Soybean Supply Stabilization Corporation. This scheme has meant that not only has the price of soybeans been kept low but they have been comparatively stable.

The expansion of the livestock sector between 1960 and 1990 has meant that ever-greater reliance has been placed on imported soybeans. The decline in the self-sufficiency ratio for soybeans was particularly dramatic between 1960 and 1970, falling from almost 30 percent to less than 5 percent (Simpson et al., 1985).

Even with the low import price, however, soya-based protein has been slowly losing ground to meat as a major protein source in the Japanese diet. Table 6.1 shows the small decline in the importance of soybeans as a source of protein. It should be pointed out that between 1960 and 1987 the Japanese daily protein intake increased from 69.5 grams to 85.5 grams. In short, soybeans will take a secondary role in Japanese consumers' future search for dietary protein.

FISH AND SHELLFISH

Fish has traditionally been the major source of animal-based protein in the Japanese diet. In the last few years shellfish have increased tremendously in popularity. This has been the result of higher incomes as well as improvements to shellfish production and processing technology.

Fish remains more important than poultry and red meat in the Japanese diet. By the end of the 1980s, average per capita consumption of fish and shellfish was 100 grams per day whereas that for meats was about 25 percent less (at 74.6 grams). Fish consumption has increased 25 percent since 1960. Hence, the prospects for beef consumption will be closely tied to trends in fish consumption and available supplies.

The preeminence of the fishing industry can be attributed to three major factors. First, Japan does not have much land suitable for extensive raising of livestock. Second, the dietary ban on the consumption of meat from four-legged animals inhibited the development of the livestock industry for decades after the ban was formally lifted in 1868. Third, the many miles of coastline, the relative closeness to the sea of almost all of Japan's populated areas, and

TABLE 6.1. Soybeans as a source of protein in the Japanese diet 1960-1990.

Year	Percent	Year	Percent
1960	11.5	1980	8.2
1965	9.7	1981	8.3
1970	10.1	1982	8.4
1971	9.9	1983	8.7
1972	9.8	1984	8.8
1973	9.6	1985	8.7
1974	9.4	1986	8.9
1975	9.4	1987	9.1
1976	8.7	1988	8.9
1977	8.2	1989	9.0
1978	8.2	1990	8.8
1979	8.2		

Source: Japan MAFF (1992).

favorable ocean currents–the warm Kuroshio from the south and the cold Oyashio from the north provide an environment where marine life thrives (Coyle, 1983)–led Japan to turn to the sea as an obvious protein source. Japan is known as a nation with a great seafaring tradition and the Japanese fishing fleet ranges far and wide in search of sources to supplement the islands' inshore fisheries.

The second World War severely damaged the Japanese fishing industry. This was one of the reasons for the near collapse of the domestic food system. The fleet was eventually rebuilt with modern equipment, and production grew rapidly until the early 1970s. The long-distance fleet grew in importance in the era of cheap energy and before the general imposition of 200-mile (320-kilometer) fishing zones in the early 1970s. Since then, the inshore fisheries have regained their previous importance. Approximately 45 percent of Japanese seafood consumption comes from local offshore and coastal fisheries. Long-distance fisheries account for 20 percent of the consumption; aquaculture accounts for a further 9 percent.

Composition of the Japanese catch has, however, been changing. Supplies of cheap fish such as sardines and mackerel have been relatively abundant while those for the table trade–crab, herring, squid, and cod–have been declining. Luxury fish such as tuna and salmon have had only small increases in their catch. The main effect has been that the price of table fish, the main competitor for live-stock products, has increased considerably, leading people to substitute meat for fish in their diet. The decline in catch has, however, been somewhat offset by an increase in imports.

The prospects for a major expansion of fish supplies appears remote. The fisheries within the Japanese 200-mile (320-kilometer) limit are being managed at near-capacity levels. Fish stocks on the high seas are being steadily depleted due to the inability to secure international cooperation to manage the fisheries. The coastal areas used for aquaculture suffer from pollution from thermal power plants, industrial waste, and agriculture.

Fish consumption in Japan is among the highest for developed economies. Consumption per capita is about five times that of North America and Australia and twice that of the United Kingdom. While total consumption of fish and shellfish has been increasing relatively slowly for some time, the composition of the fish consumption basket has been evolving. Consumption of low-priced fish has been decreasing, while consumption of high-priced fish has been increasing, reflecting rising incomes. Seafood consumption has also been helped by the increasing popularity of restaurant meals. Salmon, tuna, and flatfish have seen the greatest increase in consumption. Fish constitute about 80 percent of domestically secured supplies. The products of fish farms–primarily oysters, scallops, sea bream, and yellowtail–represent about 9 percent of domestic consumption. Molluscs such as squid, octopus, and clams represent about 6 percent of domestically produced supplies and crustaceans–lobster, crab, and shrimp–another one percent. In addition, there are about 25,000 establishments engaged in the cultivation of edible seaweed.

A large proportion of fish production is processed. About 36 percent of the fish used for human consumption is salted, dried, or smoked. Only 10 percent is canned whereas a large proportion, 25 percent, is consumed in the form of paste. Only about 30 percent is eaten in fresh or frozen form.

As with much of the Japanese food distribution system, movement of fish to consumers is a cumbersome process that involves many layers of market participants. Local markets handle almost all the fish landed in Japan. About half the fish moving through these markets goes directly to the local retailers. This ensures that a large proportion of the catch is available in a very fresh form. This increases the competitive pressure on the beef industry to provide product in a similar fresh form. Fish not sold in the local market is either purchased and shipped to other wholesale markets by dealers or processed.

Small shops dominate the retail trade for seafood although they are grudgingly giving up market share to the supermarkets. This is very similar to what has been happening in the meat distribution system and, hence, gives no advantage to seafood over beef.

The major problem for the Japanese is to maintain production from the domestic fisheries. Much of the existing management of domestic fisheries has been focused on maintaining short-run output rather than on the long-term viability of the industry through conservation. Hence, it would seem that fish production from Japan's domestic resources will, at best, remain relatively constant except for a few product lines. Considerable funds are being committed to improving aquaculture production and there is a general re-orientation toward fishing resources controlled directly by the Japanese.

If the domestic market is not going to be self-sufficient, then imports must make up the remainder of the supply reaching the consumer. Imports constituted approximately 25 percent of Japanese seafood consumption in 1990, up from 5 percent in 1970. A larger reliance on imports can only help beef consumption since seafood imports tend to be more expensive than beef imports. High-value items such as fresh or frozen shrimp, prawns, squid, and other crustaceans and molluscs, along with tuna and salmon, make up over half of Japanese fish product imports.

One major disadvantage for countries wishing to export beef to Japan is that the import tariff rates on fish and shellfish are considerably lower than those for beef. As a result, while the export price of fish may be higher than the export price of beef, when the tariff cost is added much of this differential disappears. The Japanese tariff system is complex, with different rates for different sources of supply. The rate for GATT signators tends to be below the rate for non-signators but above the preferential rates applied to the exports of a large group of developing countries under the Generalized System of Preferences. For most major beef exporters–Australia, the U.S., New Zealand, the EC, Canada–the GATT rates apply. These range from 3 to 16 percent. After the 1991-1992 adjustment period, the beef import tariffs are bound at 50 percent. Since one quarter of the fish for the premium market in Japan is imported, it would seem likely that the domestic price will reflect the imported price. Beef exporters should therefore monitor international fish prices when planning their export sales.

POULTRY

As with the poultry industry in many countries, the Japanese poultry industry has undergone significant changes in the last two decades. These changes are largely the result of technological advances in the production of poultry. It has meant the rapid expansion of chicken production and virtual self-sufficiency in this product. Since the technology of chicken production can be transferred easily from other countries and since the Japanese industry uses inexpensive imported feed, poultry production in Japan has become competitive with that of most of the world.

The modern era in Japanese poultry production began with the development of the specialized egg-laying industry in the first quarter of the twentieth century. As a result, until the end of World War II, most chicken meat came from layers that had outlived their productive lives. Rising post-war incomes brought an increased demand for chicken–first in the restaurant trade and later, by the 1960s, in the home consumption market. Since that time there has been a general consolidation of the specialized broiler industry. Chicken production is typically concentrated close to population centers. The number of farms decreased and the average size of those that remained increased. The net result was a seven-fold increase in poultry production from the middle of the 1960s to the beginning of the 1990s. This reflects the economies of scale that come with controlled environments and mechanization of feeding and cleaning systems.

The structure of the poultry production industry has also been changing, with vertical integration emerging as the dominant industrial structure. This is similar to the evolution of the industrial structure of the poultry industry in the U.S. and the EC. All aspects of the industry from consumer marketing, packaging, processing, feed supply, and chick supply to broiler raising are either directly controlled or contracted by large firms and cooperatives. Large Japanese corporations such as Mitsubishi and Mitsui are involved. As a result, both the production and the distribution systems are more modern and efficient than those found in the beef industry.

There are virtually no government programs in the poultry industry beyond those essential for health and food safety. Tariffs on chicken and poultry products are relatively low, ranging from 5 to 20 percent.

By the early 1970s, Japan was virtually self-sufficient in broiler production, imports stood at about 2 percent of consumption. As with other meat products, the self-sufficiency level has declined somewhat so that imports were approximately 17 percent of domestic consumption at the beginning of the 1990s. Imports tend to be concentrated on certain cuts, which suggests that there may be an imbalance between the mix of cuts supplied from a bird and the mix of cuts demanded by the consumer. This generally conforms to the consumption habits of Japanese housewives whose

> . . . most popular poultry cut is the low-calorie portion of the chicken breast referred to as sasami in Japanese supermarkets. Housewives' second choice tends to be the regular breast cut, followed by liver, wing, leg, thigh and gut. Uncut chicken (whole chickens) and half chicken are seldom purchased. (Khan et al., 1990, p. 47)

In other words, importation of entire chickens is unlikely because domestic supply can easily expand to fill demand.

Total available chicken supplies for the period 1965 to 1991 are presented in Table 6.2. Chicken consumption increased at an annual rate of 27 percent between 1965 and 1970. Between 1970 and 1980 the average annual rate of increase was just over 10 percent. In the 1980s, growth was less than 5 percent per annum. In 1965, total beef and chicken consumption were approximately equal (231,637 and 210,734 tons, respectively), while in 1991 chicken consumption was 60 percent greater than beef consumption (1,732,465 tonnes compared to 1,081,939 tons).

The rapid increase in chicken consumption reflects the ability of the industry to capitalize on the available technology by providing a relatively low-priced source of meat. As a result, imports remained relatively free of restrictions. In the beef industry, this type of technological advantage was unavailable to Japanese producers and the industry could not greatly increase output. To prevent a large increase in beef imports, the government imposed import quotas.

It appears that Japanese chicken consumption has reached a mature stage. This is suggested by the slowdown in the rate of increase in chicken consumption. This means that chicken will now have to compete for the consumers' meat dollar on the basis of price. Since the supply price of chicken meat will depend largely on the price of imported feed, any decrease in beef price as a result of liberalization will put beef in a more competitive position relative to chicken.

Hence, it would seem that chicken and beef will actively compete for market share in Japan in the future. This is similar to the North American situation in which the battle for market share is being fought by the same two commodities. If the rapid rate of technological change in the chicken industry can be sustained then chicken will gain market share from beef because its price will fall relative to beef. Further, if the trend toward greater consumption of processed and fast food continues in Japan, chicken may further increase its market share since, at least in North America, chicken has proved far more adaptable to product innovation than has beef.

PORK

As the other red meat in the consumer's food basket, pork is the product that substitutes most closely for beef in the marketplace.

TABLE 6.2. Total supply of chicken in Japan 1965-1991 (tons).

Year	Supply	Year	Supply
1965	210,734	1983	1,341,012
1970	499,958	1984	1,413,735
1975	758,139	1985	1,455,544
1976	860,290	1986	1,554,235
1977	947,887	1987	1,632,513
1978	1,063,508	1988	1,711,418
1979	1,160,844	1989	1,718,511
1980	1,196,265	1990	1,712,390
1981	1,223,398	1991	1,732,465
1982	1,286,358		

Source: Japan MAFF (1992).

Along with beef, pork suffered from the prohibition against eating of the flesh of four-legged animals. Hence, it has no traditional place in the diet of the Japanese. Pork producers, however, have had some distinct advantages over beef producers when supplying the domestic market. First, like poultry but unlike beef, pork production requires only small amounts of land. Second, the pork industry did not develop out of a draught animal industry and, hence, was allowed to import advanced genetic strains of hogs from offshore. Third, the Japanese taste for pork is similar to that of other nations so that the breeding stock developed in other countries could be used in Japan. Fourth, feeding regimes for Japanese hogs are similar to those in the countries which supply both Japan's breeding animals and Japan's pork imports. Fifth, since livestock feed can be imported without tariffs, Japanese pork producers can compete more readily than beef producers who are wedded to either slow-gaining domestic cattle breeds or dairy breeds not bred for efficient beef production. As a result, Japan has become approximately 80 percent self-sufficient in pork despite far less protection against imports than the beef industry has received.

While there has traditionally been a limited amount of indigenous hog production in southern Japan, it was not until the Meiji era (1868-1912) and the lifting of the ban on the consumption of four-legged animal meat that hog production began on a commercial basis. Hog-raising practices are similar to those in other modern pork-producing nations, with production taking place either in integrated farrowing to finishing operations or with the breeding and feeding undertaken by separate firms. Farm size has also been on the increase, reflecting the inherent economies of scale in pork production. In 1960, hog farms averaged only 5.6 pigs per farm. In 1990, there were about 275 pigs per farm (Japan MAFF, 1992).

Pork enters the market in either fresh or processed form. The processed product is largely smoked or cured. The fresh pork market has shown the strongest growth, but growth in higher valued processed pork items has also been strong. Supermarkets and small retail shops provide the major outlets for fresh pork used in the table trade.

About one quarter of Japan's pork supply is purchased by meat processors. Five large processors control about 80 percent of the market: Itoham, Nippon Meat, Prima Ham, Marudai Foods, and Snow Brand Foods (Khan et al., 1990). Loin ham, while cured more sweetly and with less salt than that in North America, is similar in taste to Canadian bacon but is sold in very lean form in small packages, typically 550-700 grams (Knipe and Rust, 1989). Bacon is a lower priced item. The most popular retail product is lean shoulder bacon. It is sliced at the retail counter to customer specifications and is not served as a breakfast product but rather served as the main meat dish for any meal.

Approximately 20 percent of the Japanese pork supply is imported. While average levels of protection are low compared to beef, Japan has consistently manipulated pork imports to further two objectives: to stabilize domestic price and to ensure a secure supply. Japan uses a system of variable import levies similar to that of the EC to help stabilize the domestic price. The nominal duty on pork is only five percent, which is, of course, much lower than the 50 percent that applies to beef after 1993. However, the LIPC sets a floor (and ceiling) price for domestic pork. If the price falls below the floor price then a premium is added to the tariff so that the landed import price plus the duty equals a price midway between the price ceiling and price floor. Hence, as the import price falls, the import duty rises, which helps maintain the domestic price. This mechanism may work to the advantage of beef exporters who face a fixed tariff level. If the price of beef in beef-exporting countries falls considerably and the consumer price of beef in Japan declines, through consumer substitution downward pressure will be put on Japanese pork prices. The variable import levy will then increase and, hence, limit total pork supplies by discouraging imports. This will happen while beef imports are increasing.

The Japanese have also been very careful to limit the supply received from any one exporter (Kerr, 1985). In the absence of self-sufficiency, the simplest way to provide for food security is to diversify one's sources of supply. This reduces the effect of a supply disruption from any one exporter. The Japanese have been very conscious of supply security issues since U.S. soybean exports were interrupted in 1974. Historically, pork imports have been roughly apportioned among the U.S., Denmark, and Canada. In recent years, Taiwan has become the major exporter of pork to Japan and the other exporters' share has declined. Taiwan's geographic proximity to Japan has made it possible for Taiwan to provide the Japanese market with chilled pork and to a considerable degree, this accounts for their success. The ability to supply chilled product has important ramifications for beef exporters because the beef consumer is following the pork consumer in his/her expectations regarding chilled product in the marketplace. Those who can land chilled beef in Japan are likely to see their market share increase.

Pork consumption, while growing, has been doing so at a decreasing rate. This is illustrated in Table 6.3. Between 1965 and 1975, pork consumption grew at a rate of almost 20 percent per year. Between 1975 and 1985, the rate of market growth declined to an average of five percent per year. Since then, it has fallen to less than 3 percent per year. As with chicken, pork has entered a mature stage of consumption, with relative price likely to be the major determinant of its share in the consumer's meat consumption.

TABLE 6.3. Total pork supply in Japan 1965-1991 (tons).

Year	Supply	Year	Supply
1965	407,322	1983	1,665,180
1970	758,776	1984	1,702,002
1975	1,217,514	1985	1,802,078
1976	1,268,759	1986	1,846,748
1977	1,323,890	1987	1,982,017
1978	1,432,350	1988	2,039,804
1979	1,618,666	1989	2,086,555
1980	1,630,149	1990	2,044,753
1981	1,661,196	1991	2,072,209
1982	1,629,063		

Source: Japan MAFF (1992).

INTERRELATIONSHIPS AMONG THE SUBSTITUTES

To determine the interrelationship between products economists use a measure called "cross-price" elasticity. It measures the percentage change in the quantity of a product consumed for a one percent change in the price of a substitute. If, as the discussion of the major substitutes for beef suggests, the major changes in meat consumption are likely to come from changes in their relative prices rather than, for example, changes in income, then cross-price elasticities among meat competitors become very important and provide considerable insights into changes in consumption that might be expected over time. Recent estimates of "cross-price" elasticities by Wahl et al. (1987) suggest that pork and beef are highly substitutable as price varies, while chicken and beef are much less substitutable. While the cross-price elasticity varies with the assumptions made during estimation, the expected response to a 1 percent increase in the price of pork is a 0.65-1.21 percent increase in the quantity of beef consumed in Japan. Conversely, the pork consumption response to a 1 percent increase in the price of beef is 0.29-0.90 percent. The beef consumption response to a 1 percent increase in the price of chicken is estimated to be 0.27 and 0.47. Clearly, this suggests much less substitutability between beef and chicken than between beef and pork.

SUMMARY

While the relative prices of all other protein sources will be important, it is the relative prices of beef and pork that will be the most important determinant of the proportion of the Japanese consumer's food budget spent on beef. Since the other livestock industries are relatively mature and operate with relatively low levels of protection, it should mean that there are unlikely to be any radical changes in their prices. Hence, those factors that are likely to affect the price of beef are fundamental to determining the relative consumption shares of the various livestock products in the future. This is the subject of the next section.

PART III:
BEEF PRICES IN JAPAN

In Japan the pricing of beef is very complex. The prices paid by the final consumer are determined by the complex interaction of feeding costs, government policies, prices of imported supplies, processing costs, and wholesale and retail markups.

The Price-Quality Continuum

One of the major problems that arises when outsiders first observe beef prices in Japan is that they tend to assume that the limited set of prices they observe can be taken as typical of beef prices in the entire Japanese market. They cannot. This tendency to generalize prices arises from Westerners' attempts to transfer their own experiences to the Japanese marketplace. In most developed economies, beef purchased for home consumption varies very little in quality–one or two of the top grades at most. Tastes tend to be fairly homogeneous among consumers within Western countries, although they may vary considerably among countries. Hence, while there may be price variations among cuts, these tend to be concentrated around a very narrow band and one price can be considered as a proxy for most other prices. Lower grades of beef from domestic cull dairy and beef cows, as well as from frozen manufacturing quality imports, usually do not end up on retail shelves. Rather, they are used by the restaurant trade or for food processing, commonly denoted as "hotel, restaurant, and institutional" (HRI) trade. Home consumers seldom observe these prices directly.

As a result, many foreigners casually observe retail prices in Japan, which can be as high as $500 per kg, and do a back-of-the-envelope calculation on what it would cost to deliver their domestic beef to the Japanese marketplace. They then return home with glowing reports of the large profits available in the Japanese market. Of course, these projected profits are invariably too optimistic because the Japanese prices quoted are not for the same quality of beef as is produced for the exporters' domestic markets. In fact, nowhere else in the world does one find a product similar to that which commands premium prices in Japan. The combination of tastes, cooking style, and high incomes allows for the consumption of a limited quantity of beef that is very expensive to produce. Most consumers, however, cannot consistently afford this expensive beef and not all dishes require the particular cooking characteristics of the very expensive beef. Until the liberalization resulting from the 1988 BMAA, import restrictions constrained imports to the lower-quality segments of the Japanese market. Consumers purchased imported beef because the quality beef they desired was simply not available. As a result, the beef consumed in Japan has a much wider quality spectrum than is common in most countries.

The basis of the price-quality continuum in Japan is the degree of intramuscular fat that the meat exhibits. The greater the amount of intramuscular fat, or marbling, the higher the price. Other quality factors such as tenderness, flavor, meat texture, and meat and fat color also play a role in price determination.

As discussed in Chapter 2, the abhorrence of blood, along with a culinary tradition based on chopsticks, has led to a preference for beef cooked in liquid–the well known Sukiyaki and Shabu Shabu. By slicing the beef paper-thin, no blood shows in the meat, and none will drip onto the plate. Boiling in liquid is also consistent with Shinto purification rights. Of course, the small strips can easily be handled with chopsticks. The process of boiling, however, turns muscle hard and dry. Sufficient deposits of fat ensure that the beef remains tasty and tender. To produce heavily marbled beef, however, requires a long and slow feeding period. The Japanese cattle that produce the top-priced meat may be on high-energy rations for up to 900 days (compared to 120-140 days in North America). As animals age, they tend to become less efficient converters of feed so that production costs increase rapidly as the time on feed is extended.

The largest segment of the Japanese quality continuum tends to be made up of meat from animals that are on feed for 300-400 days. While these animals are not as heavily marbled as those that produce the highest quality beef, they would still be considered extremely overfat and discounted quite heavily in exporting countries.

The LIPC specified its tenders in the quality attributes of its export suppliers. For example, the so-called "high quality beef" quota was specified so that it conformed to U.S. Choice, a very lean product by Japanese standards. As a result, U.S. beef was heavily discounted in the Japanese market. Australian and New Zealand beef has been mostly grass-fed and exhibits almost no marbling. It has tended to be priced near the bottom of the price band. Medium-quality beef tends to be used in hamburger-based foods and barbecues, whereas low-quality product is used as flavoring for soups and noodles. Since total consumption has been restricted by quotas and quality of imports regulated, the entire quality continuum has been available at the meat counters of retail outlets.

Fresh versus Frozen

Japanese consumers, along with consumers in the rest of the world, prefer fresh to frozen beef. The preference that beef be in fresh rather than frozen form is particularly important in Japan where freezers are few and the capacity of refrigerators is limited. Unlike the North American consumer who tends to buy fresh meat once a week and then freeze it in their home, Japanese housewives tend to shop daily and serve the beef the same day it was purchased. Little processing takes place in the home as butchers slice the beef into the table-ready paper-thin pieces. When beef is frozen, it does not cut readily into the desired thin strips.

Butchers often complain that beef which has been frozen and thawed does not slice well and tends to have a coarse texture. As a result, frozen beef tends to be discounted 30 to 40 percent compared to fresh beef of similar quality and it tends to be used for dishes where low-quality beef will suffice.

The LIPC specified its import tenders in frozen beef. This frozen beef was used in the LIPC's overall marketing strategy where it was used to both support the price of domestic beef and stabilize the domestic price through the strategic use of frozen storage. By allowing only frozen beef to be imported, the LIPC indirectly supported the price for domestically produced beef. Chilled (the industry name for fresh) beef, however, is a perishable commodity and must be moved through the distribution system virtually without pause. As a result, it is not suitable for holding as stocks. Frozen beef, on the other hand, can be stored for a considerable period. Hence, it is ideal for stock management. Since frozen beef and fresh beef are imperfect substitutes, when only low-quality beef was allowed into Japan, demand for Japanese beef was increased as consumers' substitutes were limited to less-desired products. Therefore, consumers were denied access to close substitutes. With the gradual phasing out of the LIPC as a result of the BMAA, imports of chilled beef increased substantially. By the end of 1989 it constituted 40 percent of beef imports (Gorman, 1990).

Since exports were restricted to frozen beef until very recently, supplying countries had no incentive to develop better chilling technology–to extend shelf-life–and allow successful ocean shipment to Japan. The chilling technology used for domestic markets in exporting countries provides a shelf-life of 40-45 days. Sea freighting for Japan requires up to 15 days, and the Japanese distribution system is slower than those in the supplying countries. As a result, a minimum of a 60-day shelf-life is required, and 90 days is preferred, by Japanese distributors. A few plants in Australia and New Zealand have reportedly achieved a 100-day shelf-life, while some plants in the U.S. can achieve a shelf-life in excess of 60 days (Savell, 1990). Being able to ship by sea is of crucial importance because air freight is several times the cost of sea freight and the tariff on beef is applied on a c.i.f. basis. This means that the cost of the air freight is added to the price used to calculate the tariff. This considerably increases the price of exported beef and in many cases may be the factor that determines whether a shipment is profitable or not (Hobbs, 1990). In the future, frozen beef will be largely confined to the institutional and processed food market.

Chapter 7

Farm Prices for Beef Animals

There are two major sources of beef in Japan: imports and domestic production. Domestic production can be further broken down into two major sources: Wagyu beef and dairy beef. Wagyu producers concentrate on supplying higher quality beef whereas the dairy herd produces a lower-quality product. The dairy herd, however, provides the greater portion of the Japanese domestic beef supply.

WAGYU

There are four recognized commercial breeds of Wagyu cattle in Japan: Japanese Black, Japanese Brown, Japanese Poll, and Japanese Shorthorn (Longworth, 1983). The Wagyu originated as draught cattle used largely in rice production. The mechanization of agriculture removed the need for Wagyu cattle for draught purposes and, as an alternative, they were bred for beef production (Kerr and Klein, 1989).

Unlike cattle in most parts of the world, Wagyu breeders selected animals for their marbling traits. Since lean meat is preferred in other high-income beef markets, breeders in supplying countries have been genetically selecting their animals to reduce marbling. As a result, they do not produce cattle with the marbling desired by Japanese consumers even when they feed their animals for comparatively long periods.

Wagyu beef production is typically undertaken on a small scale and is often a sideline enterprise on a farm. Sixty percent of the herds have only one or two animals and 99 percent of herds are composed of less than 20 animals (Kerr and Klein, 1989). Wagyu animals provide approximately 30 percent of total domestic beef production, but Wagyu producers are still very influential in the bodies that make Japanese agricultural policy.

The marbling capabilities of the Wagyu, however, are something of a myth. Certainly, some Wagyu marble very well, but the high end of the quality continuum is not consistently attained by Wagyu producers. For example, in 1984 only 1.4 percent of the Wagyu marketed reached the highest grade and just over 5 percent reached the highest two grades (Khan et al., 1990). The vast majority of Wagyu cattle qualify for only the intermediate grades which, while not as lean as the beef generally supplied by Japan's export suppliers, do have levels of marbling attainable by foreign suppliers.

The acceptance of heavily marbled beef from foreign suppliers as a perfect substitute for mid-grade Wagyu will be a major issue for those monitoring Japanese beef consumption over the next few years. A number of foreign suppliers are beginning to gear up a portion of their production to supply that segment of the market. There is some evidence that Japanese consumers do not perceive domestic and foreign beef as homogeneous products (Klein et al., 1990; Namiki, 1990). Discrimination on the basis of national origin has been observed in other markets for meat (O'Connell, 1986; Stent, 1967) as well as for a wide variety of other products (Armington, 1969; Colman and Miah, 1973). This question is of major importance for two reasons: (1) the limited opportunities for expanding Wagyu production in Japan and (2) the very high prices that Wagyu beef fetches in the marketplace. As a result, there would appear to be expanding opportunities and high potential rewards.

The small scale and subsidiary nature of Wagyu production, with the possible exception of production undertaken in Hokkaido, does not lend itself readily to expansion of production (Kerr and Cullen, 1990). Operations with one or two animals can use the forage available on small Japanese plots efficiently, but since forage is either a residual of crop production or produced on marginal land with low productivity, opportunities for expanding forage production are extremely limited. Given the relative costs and the very long feeding periods required for high-quality Wagyu production, purchasing feed is not an attractive option. Further, additional labor is not always available because many farmers hold off-farm employment. Wagyu animals are not used extensively in feedlot production since their production efficiency is less than that of dairy steers. Wagyu prices have risen considerably over the last 20years but this has not induced a significant increase in production. In short, Wagyu production has effectively stabilized and there are no apparent innovations available that would precipitate an increase in production. If demand for high-quality beef continues to expand as incomes rise, there would appear to be significant opportunities for foreign suppliers if they can (1) overcome their current production problems, (2) consistently provide a comparable product, and (3) secure product acceptance by Japanese consumers. If, on the other hand, acceptable foreign supplies of high-quality beef are not forthcoming, then Wagyu prices can be expected to rise over time.

At all stages of the beef industry, Wagyu cattle and beef tend to command a premium in the market. In the market for calves, Wagyu calves have received a considerable premium over dairy bull calves. For example, in 1987 Wagyu calves received an average price of Y408,100 per head while dairy bull calves averaged only Y202,500 (a premium of 102 percent). Between 1980 and 1987, the average annual premium of Wagyu over dairy calves was 89 percent (Namiki, 1990). In 1988, for beef leaving the slaughter plants, "full sets"[1] of 2nd grade (a relatively low grade) Wagyu steers received an average price of Y2,850 per/kg, while the same product from the carcass of 2nd grade dairy steers received Y2,022. At the wholesale level, 2nd grade striploins from dairy steers, for example, were priced at Y3,532 per kg; the same product from Wagyu animals was priced at Y5,912. This represents a premium of 67 percent in favor of Wagyu. Tenderloin fillets from the same grade of animal received Y6,358 per kg in the case of Wagyu and only Y4,320 per kg in the case of dairy animals. As a comparison, pork tenderloin sold for Y3,000 per kg and chicken breast sold for Y990 per kg. Clearly, Wagyu is the Rolls Royce of the meat trade and the standard against which other products must be compared.

DAIRY BEEF

About 70 percent of domestic Japanese beef output comes from the dairy herd. Holstein cattle form the basis of the Japanese dairy indus-

try. Since the 1960s, most of the expansion in Japanese beef production has been in the dairy beef sector. After the second World War,
largely due to an attempt to improve nutrition, the Japanese government initiated programs to encourage milk production. An inevitable
offshoot of this was an increase in the number of dairy steers available for fattening and, hence, a dairy beef industry was created. The
total Wagyu herd decreased from 2.3 million head in 1962 to approximately 1.6 million head in 1990. In contrast, the Japanese dairy herd
increased from under one million head in 1962 to nearly three million head during that time period (Hayes et al., 1990).

In order to produce the higher levels of marbling desired by
Japanese consumers, cattle must be fed a high-energy grain ration at
a slower rate of intake than in North America. Since cattle are fed
for a considerably longer period of time, slaughter weights are far
heavier than those reached by cattle in North America. Wagyu cattle
appear to be relatively better suited to this production method,
having been bred specifically for marbling characteristics. Since
dairy animals' primary role is milk production, the emphasis of
dairy cow breeding programs has been toward increasing milk
yield. And since dairy bull calves are a necessary by-product of
milk production, it is not feasible to genetically enhance their beef
production capabilities. As a result, dairy beef cattle are less genetically suited to the long feeding periods that heavily marbled cattle
require; consequently they are fed for shorter periods of time. Seldom do dairy animals reach the higher end of the Japanese beef
quality continuum.

In an effort to reap the rewards paid for higher marbling levels,
Wagyu producers have gradually been increasing the length of time
their animals are on feed. One result is increased final slaughter
weights. This is in direct contrast to developments in the North
American, Australian, and Western European beef industries where
there has been an increased emphasis on reducing the feeding period and average slaughter weights in response to consumer preferences for leaner beef. For example, in 1965, the average fattening
period for Wagyu steers was 7.3 months. Steers were approximately
20.4 months of age at slaughter, and they weighed, on average, 400
kgs. A decade and a half later, the fattening period for Wagyu steers
had risen to 18.9 months, with slaughter at the age of 28.5 months.

Dairy steer fattening underwent a similar process of change, increasing from 11.3 to 13.6 months. The average slaughter age of dairy cattle also increased, from 19.8 months in 1976 to 21.3 months in 1981, and the average slaughter weight increased by 97 kgs to 663 kgs over the same period (Longworth, 1983).

Future growth of the dairy beef industry in Japan seems doubtful. Since Japan has achieved self-sufficiency in milk production, the dairy herd will cease to grow and no additional dairy calves will become available for beef production. Beef breeds, such as the Hereford or Angus, are currently available from North America, Europe, and Australia. As yet, the beef industry in Japan has shown little interest in importing such animals as a step toward the establishment of a beef-feeding industry along North American lines. The lack of available land may be the reason for the low interest in such ventures.

Dairy beef fills the middle portion of the Japanese beef quality continuum. At the high end of the continuum one finds the top quality Wagyu product. As with beef markets in all developed countries, the bottom of the market is supplied by the third source of beef, cull cows from both the beef breeding herd and the dairy herd. Since meat tends to become tougher as the age of an animal increases, the beef from such animals tends to be of the lowest quality. Much of it goes into ground meat products or into restaurant and institutional markets where it can be tenderized. Little of this product goes into the home consumption market in an unground form.

Dairy beef takes its place between Wagyu (at one end of the quality spectrum) and beef from culled cows (at the other). Of course, there is a considerable degree of overlap between low-quality Wagyu and high-quality dairy beef. Dairy beef, however, never grades sufficiently well to receive the premiums associated with the top end of the Japanese grading system. There is also some overlap between the less preferred cuts of dairy beef carcasses and beef from cull cows.

Hence, the structure of prices at the farm gate is a function of two unidirectional forces which have their bases in the Japanese beef quality continuum. On the demand side, Japanese consumers are willing to pay the highest prices for extremely well-marbled beef. The price consumers are willing to pay declines as the degree of marbling declines. Consumers are also apparently willing to pay a premium for whatever characteristics of beef quality–whether real or perceived–that Wagyu breeding and/or production techniques impart to the meat. At the lower end of the price continuum, prices reflect other undesirable quality characteristics beyond a lack of marbling.

On the supply side, the price continuum follows the same pattern, with the lowest prices paid for production by-products and higher prices for Wagyu that reflect their higher production costs. Additional marbling requires additional feed and additional time, both of which have a cost component. Since marbling–the laying down of intramuscular fat rather than surface fat–requires a process of slow growth, there is additional cost related to extra financing costs because of the extended feeding period. Holsteins appear to have a physiological limitation to their marbling capability, which means that they are sent to slaughter with considerably less marbling than Wagyu animals and, hence, have incurred lower production costs.

GOVERNMENT AGRICULTURAL POLICIES

Japanese agricultural policy has produced one of the most highly protected agricultural industries in the world. Before the 1988 agreement to liberalize the importing system, the beef sector enjoyed the third highest rate of protection in Japan (behind only dairy products and rice). A combination of social and political factors have shaped a very protectionist agricultural policy.

The central goal of Japanese agricultural policy is to preserve the agricultural base of the economy. The ruling political party in Japan is the Liberal Democratic Party (LDP), which has been in power continuously since 1955. The political foundation of the LDP lies in Japan's rural communities. This, together with the fact that substantial numbers in Japan's urban communities still maintain close ties to the rural communities from which they migrated, gives substantial leverage to agricultural interests. As suggested in Chapter 1, farmers have a powerful political voice in the shape of the agricultural cooperatives. These cooperatives, which have close ties to the LDP, actively lobby the government on behalf of Japanese farmers.

A second motive for Japanese protectionist agricultural policies is concern over food security. The years of food shortages that followed the second World War heightened the decades-old concern of the Japanese people about Japan's limited agricultural resource base and dependency on other countries for food supplies. This fueled interest in maximizing the nation's food self-sufficiency, including self-sufficiency in beef consumption (Coyle, 1986). The intricacies of the beef-importing policy were reported in Chapter 1, but suffice it to say that the policy's major effect was to increase the prices received by beef producers.

The major legislative basis of Japanese agricultural policy is the Agricultural Basic Law of 1961. It forms the foundation of Japan's subsequent protectionist agricultural policies. This law had several objectives that are relevant to beef producers: (1) to provide opportunities for farmers to earn incomes comparable to non-farm incomes; (2) to encourage selective expansion of farm production to meet changes in consumer demand; (3) to improve farm technology; (4) to improve the marketing system; and (5) to provide price stabilization and income support. The 1961 Price Stabilization Law for Livestock Products provides the mechanisms by which beef producers' incomes can be protected from imports.

Between 1961 and 1991, Japanese policy toward the beef production sector has had four basic principles: (1) imported feed inputs should be available at competitive world prices; (2) slaughter prices should be kept above world prices; (3) slaughter prices should be relatively stable; and (4) certain segments of the price-quality continuum should be better protected than others.

Feed grains have been allowed to enter Japan relatively unencumbered. The intent of the feed grain policy is to keep costs of producing beef as low as possible so that the rate of protection required for beef producers can be kept as low as possible. However, it is doubtful that beef producers have actually been able to obtain feed at world competitive prices. The firms that handle the distribution of feed grains in Japan have considerable market power and appear to have been able to increase the cost of feed inputs to some extent (Longworth, 1983).

The major mechanism that has been used to support the slaughter price of beef has been import quotas. Imports have been restricted through quotas announced twice a year. The economic rent–the difference between the price at which foreign suppliers are willing to sell beef and the price at which it can be sold in Japan–produced by the import quota system was shared between the government, which collected a 25 percent tariff on imports, and the LIPC, which administered the majority of the quotas. The revenues obtained by the LIPC were used to promote and modernize the beef industry. By reducing the access of foreigners to the Japanese beef market through the imposition of quotas, domestic beef producers were assured of a price that was considerably higher than the world price.

Since both Japanese supply and demand for beef could vary considerably within the semi-annual quota announcements–meaning that the prices farmers received could fluctuate considerably over the period that the quota was fixed–the LIPC was empowered to use storage as a means to stabilize prices.

It is also clear that the LIPC actively supported the prices of certain segments of the industry to the detriment of other segments. The segment receiving the lion's share of support was the Wagyu sector and the higher quality portion of dairy beef production (Hobbs and Kerr, 1990). This was accomplished through the quality specifications of imported beef contracts. This meant that Wagyu and, before the SBS system, higher-quality dairy beef received no direct competition from foreign supplies. Imports were allowed to compete directly with beef from cull cows which, no doubt, negatively affected its price. As this was residual production, however, farmers would not have production-cost-based price expectations, nor would they make income from this source a significant aspect of their production-planning process. Clearly, the LIPC would have liked to have all imports in the lowest quality portion of the market. In fact, much of the early massive U.S. lobbying effort was aimed at opening the market for its somewhat higher quality grain-fed product. As a result of U.S. and Australian lobbying, in the final years of the LIPC's control of the market, a growing portion of the quota– the SBS portion–allowed for imports outside LIPC contracts. Fresh imported product began to enter the distribution system in increasing amounts; however, deregulation has caused all these pricing arrangements to disappear. The question is, what will replace them?

As discussed in Chapter 1, the Beef Market Access Agreement negotiated in 1988 has led to the replacement of beef import quotas with a 50 percent tariff and the elimination of the LIPC's administration of beef imports. As a result, the prices received by Japanese beef farmers at slaughter should, over time, come to reflect the landed (or cost, insurance, and freight–c.i.f.) price plus 50 percent tariff and will fluctuate with the changes in the international price. The LIPC can no longer specify the quality of imports or that they be in frozen form, which suggests that Wagyu and higher quality dairy beef destined for the home consumption market may encounter direct competition from imports. Hence, Japanese producers will face the possibility of lower and almost assuredly fluctuating, prices. Under normal circumstances, this would lead to a reduction in the resources committed to beef cattle production.

The Japanese government intends, however, to compensate beef producers by subsidizing feeder cattle prices. The apparent objective is to maintain the size of the industry at its existing level (Takahashi, 1990). Also, the LIPC can still intervene in the market to stabilize price. The difference under the new system is that exporters are not forced to deal with the LIPC. The LIPC is able to enter the market to purchase and store beef. Its effectiveness at being able to stabilize downward price movements is, however, very limited. Further, a storage policy can only be used for frozen product. Since the market appears to be moving toward imports of chilled beef, intervention for the purpose of stockpiling may not be possible. To the extent that frozen product substitutes for fresh product, the purchase of frozen beef can be effective. However, the storage effort has to be large relative to the reward, and the LIPC may not find that it is generally worth the effort. The work of Hobbs (1990) suggests that chilled beef produced from North American cattle fed to the lower end of the dairy beef quality level can be price-competitive with the 50 percent tariff in place.

SUMMARY

In the future, the slaughter price of beef in Japan will be determined by the actions of export suppliers to a far greater degree than it has ever been in the past. Hence, it is important to assess the likely ability of the beef industries in potential exporting countries to supply competitively-priced beef to the various segments of the beef quality continuum. This is the subject of the next chapter.

Chapter 8

The Competitors–
Exporters of Beef to Japan

There are only a limited number of countries that can be seen as potential suppliers of beef to Japan. This is because Japan is "foot-and-mouth" free and does not allow imports from countries that have the disease. As a result, the major beef-exporting countries in Latin America, Africa, and much of Europe are excluded from the market. Table 8.1 reports Japanese imports by country since 1968.

As can be seen from Table 8.1, Australia and the U.S. have been the most important foreign sources of beef, collectively supplying about 95 percent of the market in recent years. New Zealand has run a consistent but distant third. Other countries have had spotty export records at best. Liberalization may provide increased opportunities for exports to Japan from Canada and some members of the European Community. A brief examination of these suppliers' beef industries and their likely effect on Japanese farm gate prices is presented in this chapter.

The exporters do not provide a homogeneous product since available resources and consumer tastes differ considerably from country to country. Further, all exporters are now facing the realities of Japanese liberalization, with its opportunities to produce and export beef that conforms more closely to Japanese consumer preferences than was the case when the LIPC could specify the quality of imports. Exporters' ability to respond to this challenge will clearly affect future farm gate beef prices in Japan.

AUSTRALIA

There are two major systems by which beef can be produced: (1) grass-fed systems; and (2) grain-fed systems. A grass-fed system

TABLE 8.1. Japanese beef imports by country (metric tons).

Year	Australia	U.S.A.	N.Z.	Mexico	Canada	Sweden	Others	Total
1968	10,031	41	2,298	—	—	—	1,133	13,503
1969	15,062	97	3,081	—	—	—	383	18,623
1970	20,123	362	2,511	—	25	—	206	23,277
1971	36,959	507	4,004	—	22	—	79	41,571
1972	52,712	647	3,870	—	54	—	385	57,668
1973	107,271	9,527	9,464	125	314	—	523	127,224
1974	42,356	7,712	2,929	342	148	—	114	53,601
1975	37,109	3,545	3,512	18	1	—	738	44,923
1976	77,025	11,864	4,639	106	367	—	232	94,233
1977	72,055	7,330	3,903	803	284	—	172	84,547
1978	78,173	13,026	7,800	1,342	333	—	190	100,864
1979	101,268	24,672	3,510	1,447	819	—	76	131,792
1980	93,614	23,674	3,991	903	1,579	—	191	123,952
1981	87,071	27,542	6,148	773	1,959	60	94	123,647
1982	86,099	32,079	3,645	442	69	249	111	122,694
1983	91,043	37,728	7,734	17	232	133	655	137,542
1984	91,962	41,640	7,580	1,490	298	581	1,533	145,084
1985	93,129	45,938	6,965	1,602	247	944	1,382	150,207
1986	107,266	62,137	6,083	2,070	233	686	1,518	179,993
1987	120,552	82,483	7,890	4,051	296	128	1,271	216,671
1988	134,490	106,556	10,487	4,441	594	20	1,835	258,423
1989	172,336	144,357	14,010	4,301	1,808	119	2,190	339,121
1990	191,163	157,857	10,258	3,631	1,544	45	1,381	365,879

Source: Japan MAFF (1992).

uses either grazing or grass-hay silage to feed cattle over the entire production period from weaning to slaughter. A grain-fed system uses rangeland, pasture, or grass-hay primarily for the breeding phase of cattle production. At some point after calves are weaned, they are placed in a feedlot and fed a high-energy grain-based diet until they are ready for slaughter. Often these two stages of production are geographically separate and are undertaken by separate firms. Cow-calf operations specialize in breeding, while feedlots specialize in fattening. Cattle grow faster and reach slaughter weight at a much younger age on high-energy rations. Since meat tenderness declines with the age of the animal, grass-fed beef has tended to be tougher than grain-fed beef. Grass-based production does not foster marbling in animals, so carcasses tend to be extremely lean. Further, a diet of grass tends to give a yellowish tinge to beef fat. The Japanese have a strong preference for meat with white fat. As a result, beef from grass-fed production systems tends to be classified as being at the lower end of the Japanese beef quality continuum. This was the type of product the LIPC preferred to import when it set import quotas prior to 1991.

Australia has traditionally concentrated on producing grass-fed beef. As a result, this meat was the preferred import product by the LIPC. This was because the lean product from relatively old animals with a yellow fat color could compete only with the lowest valued products on the Japanese quality spectrum. Of course, it was shipped in frozen form to conform to the specifications of the LIPC contracts. This product had very little substitutability with domestic Wagyu or well-marbled dairy beef.

The product Australia shipped to Japan is the same product with which Australia has made its reputation as a major supplier of beef to the world. Traditionally, the United States, Canada, and Europe were the main markets for Australian beef, but with the formation of the European Community (EC), and particularly Britain's entry into the Community, the European market for Australian beef has been limited.

While the Japanese market for beef has been Australia's fastest growing market since the mid 1970s, in 1989 the U.S. still accounted for 47 percent of Australian beef exports, whereas Japan's share was approximately 30 percent. In the U.S. and Canada, grass-fed beef from Australia is used to augment supplies of lower quality beef, usually from cull cows (Kerr, 1987).

As can be seen from Table 8.1, Australian shipments of beef to Japan increased more than tenfold during the period of 1968 to 1987. This was accomplished without any major changes to the beef production system or the structure of the processing industry. Australia was able to continue to use the system, which had served it well in other world markets, for its expansion into the Japanese market. It had a willing partner in the LIPC, which was more than glad to have a source that would not compete directly with the products it had been set up specifically to protect.

Australia has very efficiently used its existing set of resources to produce beef for export. In southern Australia, cattle are run in conjunction with sheep and complement grain production on mixed farms. The climate in the northern part of Australia often allows cattle to be produced on rangeland. Grass yields are often low, requiring cattle to expend considerable energy in searching out forage, and thereby producing slow growth rates.

Compared to the U.S. or Canada, a large proportion–approximately 60 percent–of Australian beef production is exported. This has meant that Australia has been willing to commit considerable resources to market development and that all elements of the industry understand the importance of both marketing intelligence and promotion. Research and market promotion are independently funded by significant "check-offs"–mandatory deductions from the price received when an animal is sold for slaughter–which are matched by the government. Hence, Australia will continue its major promotional campaign to sell beef in Japan and its significant research effort to monitor developments in the Japanese market in the wake of liberalization.

The Australian beef industry realizes that the Beef Market Access Agreement will bring significant changes to the Japanese market and, thus, has been expending a considerable effort to anticipate changes and to react to them. The Australian strategy has three major thrusts: (1) retention and, if possible, expansion of its grass-fed beef market; (2) movement into grain-fed production to capitalize on access to the higher ends of the Japanese quality spectrum;

and (3) movement of increasing quantities of chilled rather than frozen beef.

Until the Japanese market has been liberalized for a considerable time, it will not be clear how the market shares of the various segments of the quality spectrum will evolve. It is clear, however, that a considerable market share will remain for grass-fed beef. The consumption of ground beef in both the home and the fast-food industry is expanding. With access to larger quantities of better marbled (and presumably cheaper) imported beef, Japanese consumers may switch away from grass-fed beef to some extent. In any event, whatever market exists for grass-fed beef will remain largely the domain of Australian product. The only likely competitor is New Zealand since Canada and the U.S. already import this product to satisfy demand for manufacturing beef.

The Australians, however, believe that the major opportunity for market growth lies in the higher marbled grain-fed market (Sheales, 1990). The Japanese business community seems to concur because there has been significant Japanese investment in both feedlots and processing facilities in Australia over the last few years. This has taken the form of both direct investment and joint ventures with Australian firms. From an almost nonexistent industry, feedlots expanded to the point where 180,000 cattle were on grain-fed diets in 1989 and projections suggest that this will rise to 270,000 by 1995 (Sheales, 1990). There is little domestic market for grain-fed meat and it is unlikely to be competitive in the North American market. Japan is clearly the target market for this product.

Some players in the Australian beef industry suggest that animals should be fed to an age of 22-30 months, with the carcass weight in the 340-385 kg range (Sheales, 1990). As with other suppliers, the Australians have little experience with feeding cattle for extended periods and there is a great deal of uncertainty regarding the production process. Since feedlotting of beef has not been traditional in Australia, considerable investment in facilities has been necessary over the last few years. Some of this investment has come from Japan. Approximately 10 percent of commercial feedlot capacity is wholly or partially controlled by the Japanese. This allows direct Japanese input into the feeding and handling process and may help Australian feeders overcome some of the problems associated with the extended feeding of cattle and the production of well-marbled beef.

The Japanese have also been investing in the Australian meat-packing industry. The investment has been in a small number of relatively large plants which account for 15 percent of the total number of cattle slaughtered and one quarter of the beef shipped to Japan. The investment comes, in part, from firms with links to the Japanese distribution system.

Vertical integration provides Australian firms with direct access to information about Japanese quality standards and consumer preferences. Australian reaction to direct Japanese investment in their industry has been mixed. While the benefits are well understood, there has been concern expressed that many of these benefits accrue to the Japanese rather than to the Australian industry. Realistically, there is probably some limit to the degree of Japanese ownership the government will accept. Still, direct access to the Japanese market through firms with a stake in the industry is likely to give the Japanese-owned firms a considerable competitive edge. The marketers of grain-fed Australian beef in Japan will, however, have the formidable task of overcoming the traditional Japanese view that Australian beef is a low-quality product.

The Australian industry has also been working extremely hard to improve its ability to ship chilled beef to Japan. At the beginning of the 1990s the Australians appear to have the world lead in shelf-life technology. By 1989, over 50 percent of the beef shipped to Japan was in chilled form. Hence, Australia is able to position its product for the Japanese home consumption market which, as described earlier, requires beef in chilled form.

In short, the Australian strategy will likely be to attempt to continue its dominance of the low-quality market while trying to expand into the well-marbled grain-fed market. If they can overcome the production problems associated with producing a well-marbled animal, they will have a strong presence in the home consumption market. However, even if they are successful in overcoming their technical problems, Australians may still have a difficult time remaining competitive as their ability to expand production without increasing cost may be quite limited (Hayes, 1990a). Their success in the premium market will also depend on the success of their major competitor, the United States.

THE UNITED STATES

At first glance, it would appear that U.S. beef is doing very well in the Japanese market. U.S. exports exhibited the fastest rates of growth in the late 1980s, and Japan is the largest beef market for the U.S. The competitiveness of U.S. beef, however, has not yet had an honest market test. This is because the growth in U.S. imports can be directly attributed to trading concessions that were obtained from the Japanese through intensive U.S. lobbying.

The LIPC's practice of contracting for tenders with the quality attributes specified for lean grass-fed Australian beef was perceived as a trade barrier by the U.S. grain-fed industry. This industry pressured the U.S. government to raise the issue forcefully with the Japanese government. As a result, the misnamed "high quality" portion of the Japanese import quota was created. In 1977, Japan imported approximately 85,000 metric tons of beef, of which the U.S. supplied only 7,500 metric tons (or less than 10 percent). The 1978 Strauss-Ushiba Agreement between the U.S. and Japan created the "high quality" portion of the quota and Japan agreed to have the "high quality" quota increased to 30,800 metric tons by 1983. This was to take place while the global import quota remained fixed. The result was reapportioning of the quotas away from Australia and toward the U.S. The U.S. supplied 27.5 percent of Japanese imports in 1983. The Strauss-Ushiba Agreement was followed by the Brock-Yamamura Agreement which specified an increase in the global quota of 36,000 tons by 1987. The "high quality" portion of this increase was 27,600 tons (Hobbs and Kerr, 1990).

It is clear that while exports of U.S. "high quality" beef have increased, the increase has been due to import quota changes resulting from political action. "High quality" beef contracts were specified by the LIPC to conform to the U.S. Choice grade which, while ranking higher on the Japanese beef quality continuum than Australian grass-fed product, is still considered to be of relatively low quality by Japanese consumers. The U.S. product has therefore not had a competitive test for two reasons. First, by specifying the contracts in U.S. Choice, beef from other grain-fed production systems was effectively prohibited. The expansion of the Australian feedlot industry had to await the decision to remove the LIPC from the exporting system. Canada, the other country in which grain-feeding technology predominates, produces a slightly leaner product than U.S. Choice for its domestic market. As a result, a large proportion of its production did not qualify for the LIPC contracts. However, it is not likely that either Canada or Australia could supply grain-fed beef in sufficient quantities to displace a significant proportion of U.S. beef in the Japanese market.

The second and more important test for U.S. beef may arise because U.S. beef has never had to compete directly with Australian grass-fed beef in the Japanese market. Since both products were limited by quotas, Japanese consumers were only allowed access to restricted quantities of each type of beef. In the wake of the BMAA, the LIPC contracts no longer exist and Japanese users are able to freely contract with all suppliers and specify any quality they deem appropriate. Only after some time with no quotas in existence can it be determined whether or not Japanese consumers will be willing to pay the premium that the cost of grain-fed production requires. Early indications are that the U.S. market share is relatively constant and the Australian market share increasing following the deregulation (Table 8.1). While it may seem unlikely that Japanese consumers will choose a lower quality product, that may depend on the relative prices of the two products in a market without quotas.

A more realistic question, however, relates to the type of grain-fed product that will be supplied in the wake of deregulation. There is no reason to believe that product tailored to domestic U.S. tastes will be preferred product in Japan. There is, of course, considerable evidence that Japanese consumers prefer a considerably more marbled product than U.S. Choice. Cattle that are graded U.S. Choice are typically fed grain for about 140 days prior to slaughter. If this feeding period were extended to approximately 280 days to cater to Japanese tastes, then an entirely different competitive situation would prevail. Any advantage provided by U.S. experience in feedlot technology might well be negated. This is because no country other than Japan has much experience with an extended feeding period.

Japanese experience will be of only limited relevance because of the genetic differences between Wagyu and North American beef cattle. The Japanese government prohibits the export of both Wagyu cattle and Wagyu semen, so they are not available to foreign suppliers. Hence, if foreign suppliers wish to compete directly for the high end of the Japanese quality continuum, they will need an intensive and long-term breeding program. As yet, even the basic genetic research has not been done to determine the factors that affect marbling ability. As a result, the very high end of the price-quality continuum is likely to remain the preserve of Wagyu beef producers for a long time to come.

It would appear that Australia has a small head start over the U.S. in tailoring beef for the Japanese market. This advantage arises from three sources: (1) Japanese direct involvement in production, (2) chilling technology, and (3) packing plant technology. As noted above, there has been considerable investment and direct involvement by the Japanese in the Australian beef-packing and feedlot industry. This cannot but help reduce the problems associated with "learning by doing" since Japanese expertise can be drawn upon directly. Further, vertical integration will ease the entry of new products into the complex Japanese marketplace. While there has been some direct Japanese investment in U.S. ranching, cattle-feeding, and packing industries, it has been on a very small scale and can only have limited impact on either the industries' production knowledge or access to Japanese markets.

The Australians have made greater progress with chilling technology and are able to provide the Japanese with a much longer shelf-life. The lead in shelf-life technology gave the Australians a considerable advantage during the transition period leading up to deregulation as more and more of the quota was made available for beef outside the LIPC tender system. Hence, the Australians had established both distribution networks and mechanisms to minimize any problems with chilled product prior to total deregulation. The U.S. was shipping some chilled product but this was, for the most part, air freighted.

The Australians have managed to extend the shelf-life of their product to the point where sea freighting is feasible. In the U.S. there is now considerable effort being expended on research into shelf-life technology (Savell, 1990). The structure of the U.S. packing industry may, however, make it difficult for firms to adapt to the requirements of chilling technology.

In recent years there has been a major rationalization of the U.S. packing industry, with both industrial concentration and significantly larger plants emerging. One of the results of larger plants has been economies of scale in processing. These economies have resulted, in part, from faster speeds on the processing lines. The current thinking about chilling technology is that success depends on a systems approach (Savell, 1990). This means that the meat must be handled in a very sanitary manner through the entire process, from pre-slaughter to final packaging. Fast line speeds, however, create situations in which it is very difficult to control some of the variables essential to successful chilling technology. Australian plants are smaller and have slower line speeds that allow increased care in the handling of meat. Slowing down the lines in the U.S. large-scale plants considerably reduces their efficiency. There are some smaller plants in the U.S., but they tend to be older, inefficient, and have greater difficulty attaining sanitary standards. As yet, the opportunities presented by Japanese liberalization have not been sufficient for the construction of any new plants dedicated to the Japanese market. For the most part, the U.S. beef industry has been an inward-looking industry, primarily concerned with supplying the domestic market; the Australian industry has been export oriented. The Japanese market has presented the U.S. beef industry with its first major export opportunity and it will take time to respond. There is little doubt that the U.S. industry will be a major player in the Japanese beef market. The real question is the degree to which the industry's success will be restricted by competition from Australia.

There is little doubt that the U.S. industry will be vying with Australia and the domestic Japanese industry for a major share in the premium market. There will also be opportunities, however, in niche markets for other countries. This is particularly true in light of Japanese concerns with the security of supply, which means that, in the absence of self-sufficiency, they may wish to rely on diversified sources of supply. Evidence from the pork industry would suggest this strategy.

NEW ZEALAND

As can be seen from Table 8.1, New Zealand has been the third most successful exporter of beef to Japan (although volumes shipped are far smaller than either the U.S. or Australia). New Zealand's beef industry is based on grass-fed technology and in the past has therefore had to compete directly with Australia for that portion of the quota not specified in U.S. Choice.

As with Australia, New Zealand's beef industry has an export orientation. There are twice as many cattle in New Zealand as there are people, and approximately three quarters of its production is exported. New Zealand's large dairy industry provides approximately one quarter of the cattle herd. The climate allows year-round grass production and, since alternative uses for land are limited, leads to a low-cost production system. It appears that New Zealand will continue to concentrate on producing grass-fed beef for the lower end of the Japanese quality continuum and not enter the competition for the premium grain-fed market. New Zealand does produce some feed grains but large-scale cattle feeding for Japan is not feasible.

The orientation of New Zealand's beef industry has been towards the North American market, particularly since access to the EC has been strictly limited. Approximately 85 percent of New Zealand's beef exports go to the U.S. or Canada. New Zealand beef is used to supplement low-quality beef in the North American market. New Zealand's share of the Japanese market remained at less than 5 percent of total imports throughout the 1980s. It does, however, represent the country's third largest export market. Under the quota system, New Zealand did particularly well in the "special quota" allotment for Okinawa, where they had a long-standing reputation as a quality supplier. The bulk of New Zealand's exports have been frozen product, but exports of chilled beef represent approximately 10 percent of the total.

In response to liberalization, New Zealand is apparently trying to pursue the low end of the Japanese quality spectrum and, in particular, to supply beef for the processed-food market. As a result, little beef from New Zealand is likely to end up in the direct home sales market.

New Zealand appears particularly interested in the manufacturing beef market (Harrison, 1990). Prior to 1991, quota allocations specified for table meat, as well as high prices, made beef a noncompetitive input to processed meats. It is in the hamburger meat market where this is particularly evident. In most of the world, hamburger meat is a beef product. While Japanese consumption of hamburger meat increased by more than 25 percent in the 1980s (Harrison, 1990), beef constitutes less than 30 percent of the meat used in patties and other formed products. Pork, chicken, mutton, and even vegetables and fish are used instead of beef. A decline in the price of beef should lead to both greater use of beef and a general expansion of the hamburger market.

New Zealand appears well positioned to take advantage of this market niche. Considerable experience has been gained from the Japanese sheep meat trade, for which New Zealand has been supplying over 40 percent of Japan's consumption. A large proportion of this sheep meat has gone into processed meats, and New Zealand has the only meat-trading organization in Japan owned by foreigners. New Zealand has therefore gained considerable importing experience (Harrison, 1990). Joint ventures have been formed with beef importers who had not received import concessions in the past.

It would appear that New Zealand will concentrate on the bottom end of the market. This is not to suggest that if the opportunity presented itself, New Zealand would not attempt to expand its market for the table meats trade (either higher quality grass-fed beef or even grain-fed beef). Competing with Australia for the manufacturing beef market, however, seems to be New Zealand's best option. This is particularly true if Australia is able to gain a considerable measure of success in the Japanese market with its grain-fed beef. This will put upward pressure on steer prices for feeders, thereby raising the cost of grass-fed animals. This could allow New Zealand to undercut Australian grass-fed beef in Japan.

Hence, it seems likely that a large portion of New Zealand's product will not compete directly for the table meat trade. If the manufacturing sector presents opportunities, that is where one can expect strong competition from New Zealand.

CANADA

One country whose industry may benefit from liberalization of the Japanese beef importing system is Canada. Canadian production systems are very similar to those for the United States. Canada's beef industry is based on a grain-fed system, and costs and production technology are competitive with the U.S.

The beef sector is very important to the Canadian agricultural industry. Approximately 100,000 farms in Canada are involved in some form of beef production. This represents approximately one third of all Canadian farms. Receipts from the red meat sector account for approximately 30 percent of total farm receipts (Boutin and Kerr, 1989). Canada has a substantial dairy herd that contributes to total beef production but, as in the U.S., total production of low-quality beef from culled dairy and beef cows is not sufficient to satisfy existing demand. The shortfall is imported, for the most part, from Australia and New Zealand.

Realistically, it is the beef industry in the Western portion of the country that stands to gain from improved access to the Japanese beef market since it is the only area with surplus production. The surplus production in the "prairie region" is based on Canada's large dryland cereals grain industry and extensive rangelands for grazing. Barley production is particularly important in the more northern and wetter areas. Barley provides the basic feed source for the western Canadian grain-fed cattle industry.

The industry is organized in a similar fashion to that of the United States, with cattle production split between a breeding industry and a feedlot industry that fattens the animal for slaughter. Since there is considerable surplus barley production, growth in the feedlot industry is not constrained. Additionally, the packing industry in western Canada has seen considerable investment over the last few years, and the slaughter plants are relatively modern. For the most part, however, the industry has not experienced the move to very large plants, as has been the case in the U.S. This would suggest that Canadian plants, like Australian plants, could adapt more easily than U.S. plants to the technical requirements of chilled beef production for Japan. Currently, packing facilities in Canada exhibit considerable excess capacity.

It is apparent that Canada has had little success in capitalizing on the growing Japanese market for imported beef (Table 8.1). Part of the problem with Canada's poor past performance can be explained by subtle differences between the beef produced by and for the domestic markets of the U.S. and Canada. Canadian tastes require a leaner product than what is produced in the U.S. As a result, on average, the premium grade of Canada is leaner than U.S. Choice. Since the LIPC contracts for "high quality" beef were specified to conform to U.S. Choice, the majority of Canadian product was excluded from the "high quality" beef contracts. Canadian beef could qualify for the remaining portion of the quota but had to compete directly with grass-fed beef produced in Australia and New Zealand. Canadian product is not price competitive in that market. This is because unmarbled beef can be produced in two ways: the slow grass-fed method or the fast grain-fed method. Grain-fed production is considerably more expensive. Young animals fed in Canadian feedlots are still in their rapid growth stage at slaughter, so a large proportion of the energy available from the grain is channelled into muscle growth. It is only after the animals mature that they begin to convert grain energy to fat at a rapid rate.

Beef produced in the U.S. has a small degree of marbling because it is fed for 20 to 30 days more than Canadian beef. Young grain-fed animals will, however, produce meat that is much more tender than older grass-fed animals with a similar lack of marbling. Unfortunately, beef cooked in the traditional Japanese manner does not receive any benefit from the tenderness associated with a youthful animal. This is particularly true in the case of frozen beef. Therefore, Canada produces a product that, for the Japanese, is often indistinguishable from grass-fed beef but much more expensive. The shipment of chilled product may, however, make it possible to differentiate Canadian product from grass-fed product in nontraditional dishes.

At this point the Canadian industry has two paths to choose from as it responds to liberalization. It could continue to attempt to market the lean beef tailored to Canadian tastes in the Japanese market or it could attempt to produce well-marbled grain-fed beef specifically for the Japanese premium market. While some efforts have been made by the private sector to investigate the possibility of producing well-marbled beef, the Canadian government has not provided the level of research support needed for following this path. As a result, there has been little or no research into animal production, chilling technology, or the marketing of well-marbled product.

In the past, Canadian efforts have concentrated on the strategy of attempting to export lean grain-fed beef to Japan. However, unless more research support is provided for improving chilling technology, along with some regulatory reform to facilitate the use of current chilling technologies, Canadian firms will have to either continue shipping frozen beef or be faced with the considerable costs associated with air freighting chilled beef. If this remains the case, Canadian product is not likely to be a viable competitor for the home consumption market.

If chilled Canadian beef can be moved to Japan, the Canadian strategy will be to carve out a niche for their product among those westernized Japanese who eat beef in dishes similar to those consumed in Canada. To implement this strategy, the Canadian Beef Export Federation has been formed and funded by industry and various levels of government. Canada Beef, which is the Federation's commercial name, provides marketing support within both Canada and Japan for Canadian firms who wish to ship beef to Japan. The success of its program will depend on the size of the "westernized" market and whether the Japanese can be convinced that Canadian beef is a superior product to U.S. beef, which will be a close substitute in the westernized market niche. Canada Beef is also attempting to expand Canada's share of the restaurant market. Early evidence, to 1990, shows some success is resulting from their efforts.

THE EUROPEAN COMMUNITY

While the U.S. and Australia have been the traditional beef suppliers to Japan, they may not have an overwhelming advantage over less traditional suppliers in the future less-regulated market. If other major beef-producing countries, such as member states of the European Community (EC), are quick to react to the opportunities presented by the liberalization of the Japanese beef market, then they could become a force in the market.

The beef sector operates within the Community's Common Agriculture Policy (CAP). The European market is protected from imports in order to support high beef prices within the Community. These high prices have induced additional production, which often exceeds Community requirements at the high prices. Surplus production is disposed of through exports that are facilitated by export subsidies. Of the 12 members of the EC, six can be considered important exporters: Germany, France, Ireland, the Netherlands, the U.K., and Denmark. In all but one year of the 1980s, the Community exported in excess of 600,000 metric tons of beef.

Japan has not been a significant market for EC beef. Germany, France, and the Netherlands are not considered to be "foot-and-mouth" free by the Japanese, so over the last few years only Ireland, Denmark, and the U.K. have been able to export to Japan. Recently, the U.K. also lost its status as a "foot-and-mouth" free country. Danish beef production is not large, so Ireland is the only likely source of beef for the Japanese market.

In the past, Irish exports have consisted of frozen manufacturing beef. Ireland follows a grass-fed production system and the beef produced is a close substitute for that produced in Australia and New Zealand. Irish exporters have been seriously investigating the Japanese market in anticipation of liberalization and they feel there may be a niche for their product.

The ability of the Community to provide large quantities of beef to the Japanese market, however, appears to be quite limited. First, a substantial subsidy would be required before EC beef could compete effectively with U.S. and Australian beef. Given the current state of chilling technology, EC product would have to be air freighted. This would further increase the required subsidy. While the EC is not averse to subsidizing exports, given the hostile international climate toward export subsidies, it seems unlikely that the Community would be willing to initiate a major subsidy program targeted toward a market that the U.S. and Australia have worked extremely hard to open for their products. Additionally, in the Andriessen-Kiren Agreement between the EC and Australia, the EC agreed not to subsidize beef exports into the traditional Asian markets of Australia and New Zealand. However, the liberalization of the Japanese market may lead to large increases in the quantities of imports into Japan. As a result of this market expansion, subsidized EC product could move into the Japanese market without affecting historic levels of exports from Australia and New Zealand (Perdikis and Hobbs, 1990). Relations with Australia and the U.S. would certainly deteriorate if the EC were to aggressively enter the Japanese market using export subsidies.

One possible bright spot for the EC could arise from the concerns of Japanese consumers who have a high degree of interest in food safety. Growth hormones are used during beef production in all the major exporting countries. However, between 1981 and 1988 the EC issued a number of directives that effectively banned the use of hormones in beef production. These regulations virtually excluded the U.S., Australia, New Zealand, and Canada from the European market. Australian and New Zealand production was sufficiently flexible to comply with the EC regulations. The U.S. and Canada have not been willing to comply. Since the EC is already hormone-free, it could have a considerable advantage in the market if Japanese consumers were to pressure their government into imposing a hormone ban.

OTHER COMPETITORS

A number of other countries have shipped limited quantities of beef to Japan. None of them seem particularly well placed to significantly expand their role in the Japanese market. Sweden has been a fairly consistent supplier of small quantities of beef to Japan. Sweden, like the EC, has surplus production as a result of domestic subsidy policies, and it has been able to export beef with the help of export subsidies. Hence, in the post-liberalization period when tariffs apply, Swedish exports could be subject to countervailing duties. In any case, Swedish surpluses are very limited in quantity.

The only other major potential suppliers are in Africa or Latin America. However, unless they are able to overcome their problems with "foot-and-mouth" disease, they cannot become players in the

Japanese market. If they were to become "foot-and-mouth" free, it would have significant ramifications for the entire international trading system for beef. The potential effect on the Japanese market is not clear.

SUMMARY

Clearly, the Japanese market for beef at the processor-wholesaler market interface is going to be extremely complex in the wake of liberalization. Along with the three major domestic sources of supply–Wagyu, dairy, cull cow beef–there will be many foreign suppliers attempting to fill a demand for beef that extends over a very wide range of quality. Whether any discernible trends in the various qualities emerge will depend on a complex set of interactions among prices. Those institutions currently in place that affect international trade and the distribution of beef to the consumer will have an effect on both the availability and the price for the final consumer. It should be obvious that the ability to supply a considerable portion of the quality spectrum will, however, exist through a combination of domestic and foreign suppliers.

Chapter 9

Costs and Risks
of Exporting Beef to Japan

When a product is to be marketed in a foreign country there may
be additional costs and risks for those wishing to enter that market.
This chapter aims to identify and evaluate the type of exporting
costs and risks that might face an exporter of beef to Japan. Over the
long term these costs must be reflected in the prices exporting firms
receive from the foreign market.

COSTS

The Effect of the **Ad Valorem** *Tariff*

One of the costs facing exporters of beef to Japan is the *ad
valorem* tariff. This is a tax levied on an imported commodity. It is
calculated as a percentage of the value of that commodity.

The Japanese have used *ad valorem* tariffs as part of their beef
importing policy since 1954 when a beef import tariff of 10 percent
was introduced. Initially, this tariff constituted the only formal barrier to beef imports in Japan. As mentioned earlier, beef import
quotas were eventually introduced in Japan, along with the Livestock Industry Promotion Corporation (LIPC). These became the
main instruments of beef import control while the tariffs became
relatively unimportant. The import tariff was raised to 25 percent in
1964 and remained at that level until April 1, 1991, when the new
rate set out in the 1988 Beef Market Access Agreement took effect.
Beginning in JFY 1993, the *ad valorem* tariff will be bound at a
level of 50 percent.

As described in earlier chapters, the unique Japanese preferences for beef may require that exporters produce a more costly, highly marbled beef product. In addition, chilled beef rather than frozen beef is likely to be the preferred product. The higher costs of exporting well-marbled beef in chilled form are compounded by the effect of the 50 percent *ad valorem* tariff because the tariff is calculated on a cost, insurance, freight (c.i.f.) basis. This means that insurance and freight costs are included in the base value upon which the tariff is calculated. Thus, the *ad valorem* tariff imposed by the Japanese government has become one of the major cost elements of exporting beef to Japan.

Sea versus Air Freight

As was explained earlier, the majority of Japanese consumers prefer chilled to frozen beef. This is reflected in the price discounts for frozen beef in the Japanese market. To be successful in the high-quality market in Japan, exporters will probably have to ship chilled beef to Japan. However, chilled beef incurs additional costs for the exporter due to special vacuum-packaging, strict temperature and sanitary control, and potential use of special storage and transportation facilities (Hobbs and Kerr, 1990).

Furthermore, chilled beef is a perishable product with a limited shelf-life. The shelf-life attainable by exporters will determine the method used to ship beef to Japan. Surface transportation is obviously much slower than transportation by air. It has been estimated that it takes between 27 and 28 days for beef from the midwestern U.S. to reach Japan via surface transportation–including two days for chilling and five to six days for customs clearance in Japan (Seim, 1990). In comparison, the transit time for beef air freighted to Japan might be only three days (allowing for loading and unloading). When processing time is added, and a period allowed for customs clearance in Japan, the resulting air freight time for beef is about 11 days from western North America. This difference in transportation times–11 days versus 28 days–could be crucial for a perishable product such as chilled beef.

Once in Japan, the meat distribution system requires chilled imported beef to have a shelf-life of between 45 and 60 days in order to assure quality and reduce spoilage losses (Hayes et al., 1990).

Thus, the normal shelf-life times required for domestically marketed beef in most exporting countries would probably not be sufficient to allow surface transportation of beef to Japan. Instead, exporters would be required to air freight their beef. This had been the main method of transportation used by North American exporters prior to liberalization of the beef market.

The higher *ad valorem* tariff, however, significantly increases exporters' costs if they air freight beef to Japan. Typically, air freight rates are far higher than surface transportation rates. A typical surface transportation rate for beef from western Canada to Japan in December of 1989 was $0.40/kg. This represents truck haulage from southern Alberta to Vancouver and ocean freight from Vancouver to Tokyo. This compares with a quoted air freight rate for the same period of $2.80/kg (i.e., approximately 700 percent higher).

The higher tariff specified in the BMAA magnifies the effects of the transportation costs because it is calculated on a c.i.f. basis. It has been suggested that, after liberalization, the tariff paid for air freighting beef to Japan may exceed the transportation costs by surface to Japan (Gorman, 1990). Preliminary research suggests that one effect of the *ad valorem* tariff may be to prohibitively reduce exporters' profits if beef is shipped via air freight (Hobbs, 1990).

It is possible that exporters could negotiate a lower air freight rate with an airline company. In such a case, they might be able to profitably export chilled beef to Japan via air. In the longterm, however, extending shelf-life to allow surface transportation would appear to be an important step toward ensuring lower exporting costs.

Carcasses versus Boxed Beef

An exporter must decide whether to send beef to Japan in carcass or boxed form. In the past, the majority of beef exported to Japan has been sent in a boxed form. In some instances boxes have included selected cuts such as loin items and other middle meats. In others, "full sets" or "cattle packs" containing most of the major cuts from the carcass have been sent.

Shipping beef in carcass form may have a number of advantages.

First, Japanese buyers are accustomed to buying beef in carcass form. Most Japanese beef is sold as carcasses by auction at the Tokyo and Osaka Central Wholesale markets and buyers inspect individual carcasses before buying. It has been suggested that Japanese buyers might be unwilling or, at best, extremely hesitant to adapt to a system of buying boxed beef rather than inspecting and purchasing on an individual carcass basis.

Trading in boxed beef rather than carcass beef may also be problematic because of differences in cutting practices. A particular cut of beef used in an exporter's domestic market may not be suitable for the Japanese market. Higher costs will be added to the final price of the product if meat cutters have to be retrained to cut beef differently. Furthermore, if Japanese consumers are unfamiliar with the uses of a different cut of beef, a promotional campaign aimed at educating these consumers may have to be launched by the exporter or by the retailer. This will further add to the final cost of the product (Hayes et al., 1990).

Alternatively, an exporter may decide to cut to Japanese specifications as part of the processing carried out in the exporting country. Again, this is likely to add to the final cost of the product. Cutting beef specifically to a Japanese buyer's standards may be more risky than simply shipping whole carcasses to Japan. In providing different and specific cuts of beef, an exporter is vulnerable to mistakes in cutting practices and to disputes with the importer as to whether or not these cuts satisfy the specifications of the importer. The risk of this type of error or dispute is substantially less in carcass beef trade.

Finally, carcass beef may be a less costly alternative than boxed beef if special production or processing methods have been employed by the exporter. For example, if an extended feeding program has been undertaken to produce heavily marbled carcasses, the exporter will have incurred substantial production costs. Furthermore, the exporter will possess a product that is suitable only for the Japanese market. To recoup these additional production costs, and to avoid selling the highly marbled beef in the domestic market at a price discount, the exporter must sell as much as possible of each beef carcass in the Japanese market. Thus, selling selected cuts, such as the striploin or the ribeye, in boxed beef form

may be unprofitable. Certainly, the margins earned by these selected cuts need to be sufficient to outweigh any losses incurred by
the remainder of the carcass when it is sold on the domestic market
for a heavily discounted price. Trading in carcasses avoids this
problem. Unfortunately, it is not technically feasible to ship chilled
beef carcasses to Japan.

One possible solution is for an exporter to ship boxed beef in the
form of cattle packs or full sets. These incorporate a large proportion of the carcass, perhaps 60 to 65 percent (Kerr et al., 1990a).
This would also reduce any waste due to special production or
processing methods. If heavily marbled carcasses were produced
specifically for the Japanese market, trade in cattle packs would
ensure that a large proportion of each carcass was sold on the
Japanese market rather than being disposed of in the exporter's
domestic market at a price discount.

Despite the potential disadvantages alluded to above, trade in
boxed beef may have cost advantages over trade in carcass beef.
The major advantage of exporting boxed beef to Japan lies in the
reduced transportation costs. Shipping whole carcasses means pay-
ing for the transportation and the *ad valorem* tariff associated with
sending excess bone and fat to Japan.

By exporting chilled boxed beef, processors may be able to extract higher margins for their product–if they are able to satisfacto-
rily cut to the specifications of the Japanese buyer. Customized
cutting to exact specification (for example, making portion-ready
steaks from strip loins) has been cited as a method of achieving the
best return in the Japanese beef market and of establishing a market
niche for one's product (Fielding, 1990).

An exporter of beef to Japan has to measure the benefits of shipping
beef in boxed form against the likely costs. A considerable amount of
research remains to be done on this aspect of exporting strategy.

RISKS

Supplying the Japanese beef market is likely to present an exporter with a number of unique risks. A good deal of this additional
risk arises from the changes introduced by the 1988 Beef Market
Access Agreement.

Prior to the BMAA, the Japanese beef market was almost completely controlled by the Livestock Industry Promotion Corporation (LIPC). The LIPC controlled the quantity and quality of the majority of imports. Furthermore, it regulated the number of market participants.

The quality specifications of the LIPC tenders ensured that beef imports conformed to the domestic specifications of the exporting country. There was no need for exporters to investigate the type of product preferred by Japanese consumers, or to significantly alter their product. Furthermore, most imports were of frozen beef, which provided some storage flexibility. The LIPC operated as a middleman in most sales, thereby distancing business ties between buyer and seller. As a result, the pre-BMAA Japanese beef market required little flexibility of exporters, provided few opportunities for new entrants and entailed little risk for those firms that were successfully entrenched within the system.

The pre-BMAA Japanese beef market, therefore, operated as what amounts to a truncated commercial vertical coordination mechanism. Vertical coordination refers to the means by which a product moves through the successive stages of production and distribution. This could be accomplished through spot markets with multiple buyers and sellers, through bilateral trading between two firms, through internal allocative mechanisms within a fully integrated firm, and so on. Mighell and Jones (1963) explained vertical coordination as:

> . . . All the ways of harmonizing the successive vertical steps, or stages, of production and marketing. Vertical coordination may be accomplished through the marketing price system, vertical integration, contracting, cooperation, or any other means, separately or in combination. There is always some kind of vertical coordination if any production takes place. (p. 4)

Vertical coordination is a very helpful framework for the explanation of the effect the BMAA is having on the risks associated with exporting to Japan. An understanding of these changes also highlights the extent of the market risks. The vertical coordination mechanism that prevails will be determined by the structure of the market. Hence, changes in the structure of the Japanese beef market, as a result of the BMAA, may lead to new forms of vertical coordination within that market.

Previously, as a result of the LIPC's role as a middleman in the Japanese beef marketing chain, consumer preferences were not allowed to influence the type of beef imported. The product characteristics of beef imports were determined by LIPC regulations rather than by market signals from consumer preferences. Under the post-BMAA system, consumer preferences shape the types of beef moving into Japan. Buyers and sellers are placed in new and different kinds of relationships with one another. Thus, a new form of vertical coordination is needed to replace the LIPC. The mechanism that eventually evolves will provide the means by which market signals will flow from consumers to producers. It will also be that mechanism by which the participants in the marketing chain can jointly minimize the costs associated with the new risks involved.

After the 1988 Beef Market Access Agreement, off-shore beef producers who try to penetrate the Japanese high-quality market face two specific requirements: (1) the long feeding of cattle, and (2) shipment of chilled product. While there may be opportunities to exploit market niches by supplying lean and/or frozen beef to Japan, the mass market for beef consumed in the home is expected to be for chilled and marbled beef. A total feeding period in excess of 300 days will likely be needed to produce beef sufficiently marbled to receive a premium price in Japan. Furthermore, a shelf-life of at least 60 days will be required for chilled beef shipped via surface freight to Japan. Five main risks arise from these new requirements and are a direct result of the changes introduced with the BMAA.

First, if the production of highly marbled beef requires that cattle be fed for at least 300 days, there will be a greater price risk. This feeding period is far longer than feeding times for grain-fed cattle sold in the domestic markets of exporters. Cattle in North America are normally fed high energy rations for 140 days or less. Thus, an additional feeding period of at least 160 days would more than double the time which cattle are on a feeding program. The time lag between when the decision is made to place the cattle on feed and when the cattle are finally ready to be marketed would increase the price risk faced by the cattle feeder. Cattle markets are typically volatile and the expected price for finished cattle at the outset of a feeding period may diverge considerably from the final realized price when the cattle are actually marketed. Increasing the feeding period makes it more difficult for beef producers to formulate a reliable expectation of the price they are likely to receive in the Japanese market.

The formation of price expectations requires two pieces of information: (1) a reliable market to which product can be delivered and from which price information can be easily accessed, and (2) ongoing economic analysis and forecasting of relevant economic variables that might influence that market's prices over the feeding period (Kerr et al., 1990a). Neither of these pieces of information are currently available for beef exporters supplying the Japanese market.

A second risk present in the Japanese beef market is the risk of exchange rate fluctuations. This adds to the price risk. A change in the exchange rate over the extended feeding program could cause actual profits to diverge considerably from the profits that were expected at the outset. The greater length of this feeding program increases the time period over which exchange rate fluctuations could occur, thereby increasing risk.[1]

A third risk facing the would-be beef exporter arises as a direct result of the nature of highly marbled beef. The beef will be specifically tailored for the tastes of Japanese consumers. It is unlikely to be suitable for other markets. Thus, by producing well-marbled beef, an off-shore beef producer is closing the door on a large proportion of its domestic market. The product will be overly fat by most Western standards and therefore likely to incur a substantial price discount if it must be sold in the home market (Lin et al., 1989; Hayes, 1989; Kerr et al., 1990a).

If the cattle reached optimum weights for sale in Japan and the Japanese price has fallen such that cattle cannot profitably be sold in the Japanese market, the off-shore producer is faced with two options: (1) continue feeding the cattle in the hope of a price recovery, or (2) sell the cattle on the domestic market. The first option could be expensive, particularly since little is known about feed conversion ratios of non-Japanese cattle on longer feeding programs. The second option means that the cattle would be sold at considerable price discount since they are overly fat for the domestic market.

The market-specific and thus market-dependent nature of well-marbled beef thus leaves exporters exposed to considerable additional risk. The vertical coordination mechanism(s) that evolve in the Japanese market need to deal directly with this risk.

A fourth risk facing an exporter of beef to Japan arises as a result of the beef being in chilled form. It has already been established that Japanese consumers prefer fresh or chilled beef over frozen beef. While frozen beef's price is discounted in the Japanese marketplace, frozen beef does allow off-shore beef suppliers some storage flexibility. If Japanese prices are low, frozen beef can be stored until prices recover. Chilled beef is a perishable product that lacks this flexibility. It must be moved rapidly through the distribution and marketing chain regardless of the offer price. Supplying a chilled beef market in Japan will leave an exporter extremely vulnerable to price fluctuations in that market.

A fifth risk facing beef exporters lies in the complex and intricate nature of the Japanese beef distribution system. The LIPC-dominated distribution system for imports, although extremely restrictive, was relatively straightforward. Only 36 companies were authorized to import beef. With the abolition of the LIPC role in the Japanese beef importing system, exporters must establish their own business links with Japanese firms. The Japanese beef distribution system has traditionally been a multi-tiered system with many middlemen, each able to earn a reasonable profit margin from the transfer of beef among these tiers. Exporters must now find their own way through a complex maze of importing firms, wholesalers, distributors, and retailers. Whether this is accomplished through private bilateral deals or through a more formal arrangement involving some vertical integration will depend on the relative costs and benefits of these options for both parties.

RISK MANAGEMENT ALTERNATIVES

A number of alternative mechanisms for managing those risks that arise as a result of the BMAA are considered in this section.

These include speculation, forward contracts, vertical integration, and joint ventures.[2] For illustrative purposes, the situation of a foreign processor contracting for the extended feeding of cattle for sale to Japan is explored. It matters little where in the production-processing-marketing chain these risks are studied, since the nature of the risks will not differ regardless of who in the vertical coordination process must incur the risks (Kerr et al., 1990b).

For simplicity, it is assumed that the cattle have been fed for approximately 130 days in the domestic market. A processor wishing to export beef to Japan must then decide whether to contract with the cattle feeder to extend the feeding period for an additional 160 days or buy the cattle for sale on the domestic market.[3] It is assumed that the cattle feeder incurs none of the marketing risks associated with the extended feeding period.

Speculation

This is perhaps the most basic of the alternatives open to the market participants. Cattle could be placed on a feeding program without a prearranged buyer. When the cattle are ready for market they could be sold at the going market price on the spot market. Processors who speculatively long-feed cattle in this manner are subject to a number of the risks outlined above. They run the risk that the price in the Japanese market will fall below that used as a planning price, leaving them with losses. They are open to the risk of exchange rate fluctuations. The risk from producing a market-specific product is particularly acute for a processor using the spot market with no guarantee of a buyer or a market.

Although processors may stand to make considerable profits if market conditions remain in their favor, this is at the expense of considerable risk exposure. Ultimately, planning prices are likely to be too inaccurate, cattle markets too volatile, and the risk of losses too high to induce foreign processors to speculatively feed cattle for the Japanese well-marbled beef market.

Forward Contracts

A forward contract is a formal agreement between two market participants, in this case a processor and a Japanese buyer. The buyer and seller agree to trade a specified quantity of a product of an agreed quality at a future date and at an agreed price.

Forward contracts bring more certainty than speculative long-feeding. The processor is assured of the revenue that will be received and the buyer is certain of a supply of the product at a pre-set price. The planning price used at the time of the production decision is the price agreed upon in the contract with the buyer. Thus, expected market prices equal actual market prices. This makes the production decision less risky and the market outcome more transparent. In addition, a market outlet for the product is arranged from the time the production decision is made.

Two major risks remain when forward contracts are used. These arise from the market-specific nature of well-marbled beef and the fact that it is perishable. In order to induce the necessary resource investment for producing well-marbled beef, processors may require a substantial price premium from the Japanese importer. By agreeing to a substantial price premium, the importer assumes most of the risks. The absence of an increase in the use of forward contracts since the liberalization of the Japanese beef market suggests that the risks remain unacceptably high.

The use of forward contracts in the vertical coordination process shifts trade from a spot market with numerous buyers and sellers to a situation of bilateral private contracting. This creates a substantial degree of dependence between the two parties and leaves both parties open to the practice of "post-contractual opportunistic behavior" (PCOB) or "opportunistic recontracting" by the other party.

Post-contractual opportunistic behavior is the "purposeful non-honoring of a contractual agreement for economic gain" (Kerr et al., 1990b).[4] In other words, there is a risk that one party will renege on the contractual agreement.

There is a danger that PCOB will be practiced if "appropriable specialized quasi-rents" exist. In other words, if one party has made an investment in a specialized asset–such as well-marbled beef–then that party is vulnerable to exploitation by their contracting partner who may renege on the contract. Klein et al. (1978) lay out the circumstances in which this type of behavior is likely to occur:

After a specific investment is made such quasi rents are created, the possibility of opportunistic behavior is very real. . . . The crucial assumption . . . is that, as assets become more specific and more appropriable quasi rents are created (and therefore the possible gains from opportunistic behavior increases . . .). The quasi-rent value of the asset is the excess of its value over its salvage value; that is, its value in its next best use to another renter. The potentially appropriable specialized portion of the quasi rent is that portion, if any, in excess of its value to the second highest-valuing user. (p. 298)

In the context of the market for well-marbled beef, "appropriable specialized quasi-rents" are created by investing in a costly extended feeding program. At the completion of this extended feeding program, the cattle are suitable for sale only in Japan. The next best use for the cattle would probably be in the domestic market, where it is likely they will be sold at a considerable price discount. Thus, for well-marbled beef, the potential appropriable specialized portion of the quasi rent is extremely high, since it represents the difference between the discounted domestic price for long-fed cattle and the potential Japanese market price for these cattle. There is, therefore, the risk that one contracting party will practice opportunistic behavior in order to capture these rents.

The practice of PCOB may arise because of the volatility of cattle markets and the long period between the time the production decision must be made and the time the beef is marketed in Japan. For example, suppose a forward contract has been made between a Japanese buyer and a foreign processor for the supply of a specified quantity of well-marbled beef at a specified price at some date in the future. Further, suppose that, during this time period, the market price in Japan falls. The Japanese importer may then stand to take a considerable loss. One way to reduce the threat of this loss is made possible by the existence of appropriable specialized quasi-rents: The contract may be reneged upon. The importer could claim that the imported beef does not satisfy the quality specifications of the contract and then agree to take delivery of the beef only at a price discount.

Faced with this situation, the foreign processor has few options.

Given the considerable stigma attached to litigation in Japanese business (Wright, 1979), taking the importer to court could harm future dealings with other Japanese firms. Selling the beef in the exporter's domestic market would probably not provide a viable solution since the beef would be overly fat for the domestic market. Furthermore, the beef would have to be returned to the domestic market, incurring additional transportation cost and an additional marketing delay which adds to the risk of product deterioration. An alternate Japanese buyer would be unlikely to offer a substantial premium over the discounted offer of the original buyer in a depressed market. Furthermore, the time lag required to arrange an alternate buyer adds to the risk of product deterioration.

Having eliminated these options, the processor would likely be faced with little choice but to accept the discounted offer of the original buyer. Thus, the apparent security provided by the use of a forward contract may be misleading. The market-specific and perishable nature of well-marbled beef creates the necessary conditions for the practice of opportunistic behavior.

Vertical Integration

Integration is the ". . . expansion of firms by consolidating additional marketing functions and activities under a single management. . . . [Vertical integration] occurs when a firm combines activities unlike those it currently performs but that are related to them in the sequence of marketing activities" (Kohls and Uhl, 1990, p. 215). Forward vertical integration occurs when a firm undertakes a downstream activity in the production-marketing process. In the market for Japanese beef, this might involve an off-shore processor integrating forward into the Japanese distribution and marketing sector by purchasing a Japanese retailing firm. Backward vertical integration involves a firm undertaking an upstream activity in the production-marketing process. A Japanese retail or distribution firm might purchase beef-production or processing facilities in an exporting country. Firms integrate in an attempt to organize and control the marketing chain in order to increase efficiency, gain greater power over the buying or selling process, or reduce contracting costs.

Vertical integration may emerge as the prevailing vertical coordination mechanism in a market if the costs involved with contracting rise more than the costs of vertical integration (Klein et al., 1978). In the market for well-marbled beef in Japan, a processor may be aware of the danger of PCOB being practiced by the Japanese importer but unsure of the probability of it occurring. The processor will therefore require a significant price premium because of this uncertainty. The importer, however, may be unwilling to pay a large premium to the processor. If the premium required by the processor is unacceptably large for the importer, it means that the costs of contracting outweigh the costs of vertical integration and vertical integration will likely occur (Kerr et al., 1990a).

Forward integration into the Japanese distribution and retail sectors by off-shore beef processors is unlikely to occur on a large scale. The complex, idiosyncratic nature of the multi-tiered Japanese distribution system is likely to deter most off-shore firms from integrating forward. Language, cultural barriers, and the differences in business practices probably make this an extremely complicated maneuver. Furthermore, little is known about the regulatory barriers that reduce the desirability of owning businesses that operate in Japan. Careful research will be required before a foreign firm embarks on a process of forward integration.

Backward integration is more likely than forward integration in the Japanese beef-marketing chain. There has been some backward integration by Japanese firms into the Australian and U.S. beef-production and processing industries. There is no evidence of any forward integration into the Japanese beef-retailing and distribution system.

Backward integration by Japanese firms carries with it a number of advantages. First, the Japanese companies have the technical knowledge necessary to facilitate extended feeding programs and to process and package the beef to Japanese standards. Second, a vertically integrated firm can provide a direct link into the Japanese distribution system and marketplace. Third, internal transfer pricing may be used to reduce the cost of the *ad valorem* import tariff for the firm.[5]

Backward vertical integration does, however, have disadvantages. The most significant disadvantage for the exporting country is that, as a reward for incurring all the risks associated with producing and marketing well-marbled beef, the vertically integrated Japanese firm would also reap all the profits. Second, there may be political opposition to foreign ownership of beef-production and processing facilities. The family-farm bias of government policies in North America suggest this would be the case (Kerr and Hobbs, 1989). In Australia, moves to "curtail the so-called Japanese invasion of the nation's beef industry" (Owen, 1989, p. 3) have already been made, with political pressure being exerted to tighten the Australian Foreign Investment Review Board's guidelines.

In summary, vertical integration would indeed avoid many of the risks associated with supplying the Japanese beef market and would eliminate the risk of post-contractual opportunistic behavior. The benefits to such a change in market structure, however, would accrue largely to Japanese firms rather than to beef industries in exporting countries.

Joint Ventures

A joint venture between an off-shore beef processor and a Japanese distributing or retailing firm would enable both parties to retain a proportion of the profits while sharing risk and managerial responsibilities. Joint ventures have several advantages over a vertically integrated firm. First, as with a vertically integrated firm, a joint venture agreement provides an off-shore processor with a direct channel into the Japanese market. Since a joint venture is an arrangement between two firms, there are no foreign ownership concerns. Furthermore, a proportion of the profits remain in the beef-producing country. Since both partners share the profits, they both have an incentive to minimize risks throughout the production-marketing process. Finally, a joint venture partnership may have significant cost advantages since the arrangement can utilize the experience of both partners (Kerr et al., 1990a).

A joint venture can also have disadvantages. This is particularly true for a situation in which substantial risks exist and, hence, must be apportioned between the joint venture partners. The major disadvantage of joint ventures is the potential for managerial disagreements between the two partners. Some of the problems that face joint ventures have been outlined by Gomez-Casseres (1987):

Joint ventures are said to allow firms to share information, resources, markets, and risks, to build trust among firms, to yield economies of scale, and so on. But managers often stress the costs of joint ventures, such as potential for disagreements among partners, for diffusion of proprietary information, and for creation of future competitors. (p. 97)

In the export market for well-marbled beef, the motive for forming the joint venture in the first place may contain the seeds of the partnership's eventual destruction. Whichever partner assumes the larger proportion of the risk will also likely demand a major decision-making role and a larger proportion of the profits. The division of risks and profits can be the cause for much discontent between two partners. Joint ventures between Western firms and Japanese firms also have to cope with substantial cultural differences that may manifest themselves in different approaches to negotiations and other business practices (Hobbs, 1990). Any firm contemplating forming a joint venture partnership will need to carefully weigh the advantage of such an arrangement against the potential difficulties and disagreements.

SUMMARY

There are substantial risks involved in exporting beef to Japan. These risks include (1) the price risk associated with the long lag between the time when the decision to produce takes place and the time when the product is sold, (2) changes to the exchange rate over the extended feeding period, (3) exposure to contract abrogation, (4) perishability, and (5) the difficulty of selecting appropriate customers. Alternative methods of attenuating these risks include forward contracts, vertical integration, and joint ventures.

Chapter 10

Distribution and Marketing Costs
in Japan

THE DISTRIBUTION SYSTEM FOR BEEF

The beef distribution system in Japan is a complex, multi-layered system. This, together with the existence of many middlemen in the traditional distribution channels, means that the cost of distributing beef once it is inside Japan can significantly reduce the final price realized by the exporter.

Moving Domestic Beef from Farm
to Consumer

Three main channels exist for the distribution of domestically produced beef in Japan: (1) the traditional channel, based on village-level livestock dealers and controlled largely by the butchers' guild, (2) central wholesale auction markets, and (3) meat processors or meat centers (Longworth, 1983).

Before the introduction of wholesale meat markets, meat was distributed through the traditional channels. The village livestock dealer had a virtual monopoly on all live cattle sales, purchasing cattle "on-the-hoof" from the producers and selling to large livestock dealers or meat wholesalers. These larger dealers then slaughtered the cattle, selling the carcasses to secondary wholesalers. The secondary wholesalers, in turn, sold the meat to smaller wholesalers who sold it to the specialist retail shops. There were many levels and many middlemen extracting marketing margins in the traditional meat distribution system. The system was endemic with noncompetitive price fixing and substantial barriers to new entrants. Approximately 10 percent of domestic meat production is still distributed through this traditional channel.

In 1923, the Central Wholesale Market Act was passed. The premise of this legislation was that competitively determined prices provide better market signals to farmers. It empowered local authorities to set up public, wholesale, perishable goods markets in nearby cities. The intent was that competitive forces would establish prices and the entry of new firms would be facilitated. The first central wholesale market for meat was eventually opened in Osaka in 1958. A number of central and sub-central wholesale markets were eventually established for beef.

Producers deliver animals directly to the central wholesale markets where their carcasses are auctioned. Approximately 25-30 percent of all domestically produced beef have traditionally been sold through the central wholesale market system. The wholesale markets collect a small commission on all sales (Longworth, 1983).

As a third option, Japanese farmers can sell directly to meat-processing companies and wholesalers. Prices are negotiated on the basis of official carcass auction quotations. Approximately 60-65 percent of domestically produced meat is distributed in this manner, including much of the trade in boxed beef.

Meat centers (or Shoku niku) are the newest aspect of the beef distribution channel. These new meat centers operate as fully integrated units and are responsible for slaughtering, boning, cutting, and packaging. Individual processing companies rent space in the meat centers. The Japanese Ministry of Agriculture, Forestry and Fisheries (MAFF) hoped that the new meat centers would lead to more efficient and lower-cost marketing. In bypassing the traditional multi-layered and expensive wholesaling and distribution network, it was hoped that beef would get from the "farm-gate" to the "consumer plate" at lower cost (Longworth, 1983).

Imported Beef in the Japanese Distribution System

Prior to liberalization, the distribution channels for imported beef were rigidly controlled. Only 36 designated trading companies were allowed to import beef. End-users requiring supplies of imported beef had to deal with these companies. The majority of imported beef was purchased and sold by the LIPC directly. A limited amount of beef was allowed into Japan through other LIPC channels, such as the simultaneous buy-sell system (SBS) through which end-users could deal directly with foreign suppliers. Even these purchases had to move through the designated importing firms. Imported beef was then channeled into the regular distribution system. The many levels in the distribution network meant that the price of imported beef included substantial markups by the time it reached the consumer. It also made it almost impossible for exporters to follow their product through to the final consumer. This meant that they had no control over the form in which their product reached the consumer, nor any direct feedback from consumers regarding perception of their product.

The liberalization of the Japanese beef market changes the methods by which beef is imported and distributed. Exporters are no longer restricted to dealing with the 36 designated trading companies. Liberalization, however, also brings increased uncertainty for many exporters. Previously, exporters had no control over their product once it reached the Japanese distribution system. Thus, they did not have to learn anything of the intricacies of this system. Now, however, a knowledge of the distribution system may be essential.

Despite the fact that the effective monopoly of the designated trading companies over the importation of beef has been removed, the Japanese distribution system may not necessarily become more competitive. As Kerr and Klein (1989) warned:

> While the tariffs will be increased, there will be a margin between the import price and the domestic price which other actors in the system will attempt to maintain and capture. One should never perceive the Japanese distribution system as competitive and it may be possible for segments of it to secure much of the margin by buying at a lower import price while maintaining the final sale price. (p. 37)

In order to benefit from the liberalization of the importing system brought by the BMAA, exporters may have to ensure that their products are supplied to the Japanese consumer without having to pass through numerous intermediaries. If they fail to do this, final retail prices of imported beef may not fall and many of the gains from liberalization will be swallowed up by the marketing margins extracted at various stages of the Japanese distribution system.

In the distribution system, market concentration may also limit the benefits received by foreigners from the liberalization of the market. Longworth (1983) reported that the distribution of fresh beef is dominated by five major companies. They control much of the distribution system and have financial control over 60 percent of the thousands of small meat retailers. As a result, exporters may face significant structural impediments to the successful marketing of their beef in Japan.

These impediments have been the subject of the Structural Impediments Initiative (SII) talks between the U.S. and Japan. The U.S. has been encouraging Japan to reduce many of the non-tariff barriers that are seen to exist in Japanese markets. In particular, the U.S. wanted the Daiten-Ho Law (Large Retailers Establishment Law) altered to remove barriers to establishment of large new enterprises at the retail level (Mori and Lin, 1990). This law meant that smaller retailers could refuse to allow a large supermarket to operate in their locality. These provisions have recently been liberalized as a result of the SII discussions (Hayes, 1990c).

At the retail end of the distribution chain, the traditional dominance of the small specialty meat shops in the beef trade has been eroded by supermarkets. Mori and Lin (1990) quoted the result of a December 1986 survey[1] which showed that, of 2,000 households surveyed, 258 (12.9 percent) had purchased beef in the previous month. Of these purchases, 34.5 percent were at specialty butcher shops and 48.1 percent were at supermarkets. In December of 1988, of 2,000 households surveyed, 631 (31.6 percent) had purchased beef in the last month. Of these purchases, 18.9 percent were at specialty butcher shops and 66.6 percent at supermarkets.

As of 1990, little imported beef has found its way into the specialty meat shops. The majority of imported beef at the retail level is sold in large supermarkets (Khan et al., 1990). If changes in Japanese regulations facilitate the growth of large supermarkets, this should benefit suppliers of imported beef. The distribution costs facing an exporter could fall substantially if the growth of large supermarkets means direct trade between the exporter and the end-user retail store. Smaller retail shops are often unable to cope with large export shipments:

> Many of the small meat retailers are incapable of handling typical North American size beef cartons hauled in on 48-foot trailers. I've seen retailers delivering two boxes of strip loins on a moped. You cannot simply drop 10,000 pounds at a time, it just doesn't work. (Stelfox, 1990, p. 156)

Supermarket chains have no difficulty handling larger loads. If it becomes easier for supermarkets to enter the retail market sector in Japan, beef exporters may have greater opportunities for a link with a Japanese supermarket, whether it be through contracting, joint venture arrangement, or vertical integration:

> The answer to selling is to find a compatible partner to work on the distribution. (Stelfox, 1990, p. 156)

This could significantly reduce the costs of distribution.

New institutions may also arise in response to the liberalization of imports. Recently a group of local meat distributors in Nagoya, the Aichi Meat Local Wholesale Market (AMLWM), announced plans to open a new meat center. It is known as the Chubu meat trading center, and it will handle imported beef and other meats using a computerized trading system. Japanese meat distributors will be able to place a buy order on their computer terminals and monitor offers from beef exporters. The center's aim is to provide a means of competitively setting prices for imported beef while enabling a degree of risk protection for exporters since contracts are made in advance of shipment. As noted in the previous chapters, forward contracts are open to the possibility of post-contractual opportunistic behavior. However, if these contracts were to be part of a formal market mechanism, such as a meat-trading center, there may be opportunities to design an arbitration mechanism or set of trading rules within the center which deter the practice. The creation of new institutions may be an important development for exporters as they may provide improved access to the Japanese beef distribution system.

There still exist many unknowns in the Japanese beef distribution system. Exporters will be forced to commit considerable time and resources to improving their understanding of the distribution system. Clearly, the marketing channels by which beef moves to market will partially determine both its price and market share. Aggressive marketing campaigns can, however, do much to alter the final result, but these activities are not costless and must be factored into the final price of beef.

MARKETING COSTS

Beef is marketed much less like a generic commodity in Japan than it is in the rest of the developed world. In Japan, selling beef that is identified as a particular brand is common. If beef is perceived by consumers as a product that can be differentiated, then formal marketing efforts will become an important part of success in the marketplace; however,

> Japan is one of the world's most difficult markets to sell, but it is also one of the most profitable for those who succeed. (De Mente, 1988, p. 198)

Selling imported beef in the Japanese market is likely to require a considerable marketing campaign. This will have two major components: (1) market research into the types of product required, the likely end-consumer, and the most effective type of market promotion; and (2) the promotion of the product.

Acquisition of Market Intelligence

There is little readily available market information in Japan for the foreign exporter. The Japanese government provides a small amount of price data at the wholesale and retail levels; Japanese companies tend to produce glossy brochures that do not provide a great deal of statistical data. Therefore, to learn about a market in Japan, exporters are faced with collecting their own data or, where possible, purchasing it from Japanese marketing firms. Obtaining accurate information about the Japanese beef market, however, can be an extremely difficult and expensive task. Accurate retail price data are not readily obtainable. Regular visits to Japan and the establishment of contacts in the Japanese meat trade may be essential. There are a number of possible ways to gain market intelligence about the beef market in Japan.

First, exporters could attempt to carry out their own market research. However, this may not be feasible on a large scale. In addition to the obvious language barrier, some knowledge of Japanese society and culture and preliminary knowledge of the target market will be essential. Therefore, the help of Japanese nationals on the ground in Japan may be required.

A second way of obtaining market intelligence might be through the services of a Japanese market research agency. In 1987 there were approximately 100 market research agencies in Japan[2] (De Mente, 1988). A survey of the Japanese Market Research Association found that its 43 members earned Y38 billion from market research in 1986.

The choice of a market research agency should be made carefully. Criticisms of Japanese market research include the charge that much of the research is geared towards showing impressions rather than clear facts. Often the market research staff are left to use their "Japanese knowledge" and intuition to make decisions (De Mente, 1988). Many Japanese market research agencies are extremely small–less than four employees. If large, comprehensive market surveys are required, an exporter will probably have to use one of the larger Japanese market research institutions.

It has been suggested that market research projects in Japan usually cost 10-20 percent less than they would in the U.S. (De Mente, 1988). However, given the diversity and complexity of the Japanese beef market, particularly to an outsider, the extent of market research required may be far greater than would be the case in the U.S. or other Western markets. Therefore, exporters should be careful to take full account of the costs of gaining intelligence about the Japanese market since it could be an essential prerequisite to obtaining a share of the Japanese beef market.

One alternative to "going-it-alone" or using a Japanese market research agency would be for exporting firms to pool their market research efforts. In this respect, the establishment of meat export offices in Japan by the U.S. Meat Export Federation, the Australian Meat and Livestock Commission, and the Canadian Beef Export Federation could be a useful base from which to collect market intelligence.

Promotion

The second set of marketing costs facing a beef exporter are the costs of product promotion. The Japanese consumer is bombarded by advertising via television, radio, newspapers, magazines, the walls of public transport, outdoor signs, and the mail. Many consumer goods markets are characterized by a high degree of brand proliferation. Japanese consumers have been noted for their tendency to be both brand-conscious and brand-loyal. Thus, it is important for beef exporters to establish and maintain an identity for their product. This may involve introducing branded beef and advertising through a variety of media.

Before undertaking a promotional campaign, exporters should be fully aware of the final marketplace for their products. Imported beef, particularly under the LIPC system, often ended up in institutional outlets such as hotels and restaurants. If this is likely to continue, it would be a waste of advertising resources to mount a promotional campaign through newspapers and television commercials aimed at retail consumers. Instead, promotional efforts should be directed at the institutional trade, attempting to encourage hotels and restaurants or their wholesalers to use imported beef.

If exporters are aiming for the potentially lucrative Japanese retail beef market, then promotional efforts should be aimed directly at final consumers. If this is the market to be tapped then imported beef must be made to appear as an easily identifiable and differentiated product. In the past, U.S. beef has been largely indistinguishable from Australian beef in the retail stores, having been identified only as "imported beef." If the consumer cannot identify the product as U.S. beef, then campaigns to promote "U.S. beef" would be unlikely to succeed.[3] In the last few years, the major exporting nations have attempted to differentiate their products.

Australia has taken the initiative in developing a brand image for its beef in Japan with the promotion of "Aussie Country Beef" or "Aussie Beef" for short. In 1988-89 the Australian industry spent

US$6.5 million promoting Aussie beef in Japan. This rose to US$7.7 million in 1989-90 and to US$15 million in 1990-91.

Australian promotional activities are the responsibility of the Australian Meat and Livestock Commission. In 1990, a slaughter levy of $A8.60 (US$6.60) per head on adult cattle and $A3.10 (US$2.40) per head on 40-90 kg calves was used to fund promotional activities for domestic and export markets. There has been no government contribution to promotional funds for the beef industry. In 1990-91, total levies collected for promotion were expected to be around US$50 million. About half of the beef promotion budget in 1988-89 was spent domestically, with the rest spent on export promotion (Sheales, 1990).

The Australian promotional campaign in Japan stresses product characteristics such as low fat, low cholesterol, tenderness, deliciousness, and the fact that Australian beef is chilled. In addition, the campaign emphasizes the "natural" way in which beef is produced and Australia's environment. An Aussie logo has been developed, Aussie fairs and sweepstakes have been held, and trips to Australia have been awarded.

The U.S. Meat Export Federation has also launched promotional campaigns in Japan. Up to US$20 million per year has been spent on its promotional activities in the Japanese beef market (Fielding, 1990). The Target Export Assistance program provided funding for participation in food fairs, advertisements in the media, cooking seminars, and the production of cookbooks. The U.S. promotional campaign emphasized the quality of U.S. beef, stressing that it was grain-fed and presenting it as tender, delicious, healthy, lean, and refreshing. The campaign aimed to increase consumer awareness of U.S. beef, to create a brand image for U.S. beef to distinguish it from other imported and domestic beef, and to educate Japanese consumers about the uses and cooking methods for U.S. beef.

There are four main advertising media in Japan: newspapers, magazines, television, and radio. Others that are less important include direct mail advertising, outdoor signs, transit advertising, and in-store or point-of-sale promotions. There are approximately 180 major newspapers in Japan. The three largest have a daily circulation of over 23 million, with the other papers reaching at least another 40 million people. The penetration rate of newspapers is one of the highest in the world, reaching approximately 570 per 1000 people (Dentsu Japan, 1989). Japan's extremely high literacy rate of 99.7 percent is an important reason for the high penetration rate. The home delivery system is well established. Newspapers carry considerable credibility in Japan, making them important in the formation of public opinion and a powerful advertising medium. In 1987, expenditures on newspaper advertising in Japan were Y988.2 billion. Newspaper advertising space is scarce and expensive; however, it is one of the most popular methods of advertising in Japan. Examples of 1988 newspaper advertising rates for full-page, morning edition advertisements are shown in Table 10.1.

There are approximately 100 important weekly magazines in Japan with a total circulation of one billion, plus another 2,000 monthly magazines with a total circulation of two billion (De Mente, 1988). Advertising expenditures in magazines during 1987 were Y257.7 billion (Dentsu Japan 1989). The most expensive place for a magazine advertisement is the back cover, followed by the inside front cover, then the color gravure page, then the inside back cover and, finally, an inside page. Table 10.2 provides examples of advertising rates for magazine back covers.

Television advertising is another popular, but expensive, form of advertising. The density of television commercial spots in Tokyo is three times that of New York. There are over 100 commercial television stations in Japan, with nine stations in Tokyo and Osaka alone. In addition, there are 1,500 UHF television stations and 9,450 UHF radio stations in Japan (De Mente, 1988). Food and beverage advertising comprises 25 percent of all television commercials. In 1987, Y1,174.5 billion was spent on television advertising (Dentsu Japan, 1989).

Japanese television advertisements have been described as moody and "off the wall" compared to those on North American television. Japanese observers have commented that U.S. commercials contain far too much verbal content for the Japanese viewer who does not like to be bombarded with a stream of words (Dentsu Japan, 1983). The hard-sell approach of television advertising in most Western countries would probably be unsuitable for the Japanese market. Japanese commercials take a far more visual approach, stressing human values, peace, quiet, and a return to nature. Viewed

TABLE 10.1. Newspaper advertising rates: 1988
(Yen per full page advertisement–morning edition).

<u>National</u>

Yomiuri Shimbun	41,550,000
Asahi Shimbun	35,610,000
Seikyo Shimbun	12,127,500

<u>Prefectural</u>

Hokkaido:	
Hokkaido Times	2,772,000
Tohoku:	
Akita Sakigake	2,475,000
Kinki:	
Ise Shimbun	1,950,000
Okinawa:	
Okinawa Times	2,325,000

<u>Sports Newspapers</u>

Nikkan Sports	5,629,500
Fukunichi Sports	924,000

Source: Dentsu Japan (1989), pp. 307-308.

by Westerners, these commercials often do not appear to be at all connected with the product being advertised (De Mente, 1988).

There are two types of television advertising in Japan: program sponsorship advertising and spot commercials. For the first type, in addition to the cost of producing the commercial, there is a time charge and a network fee when the program is aired on a network. The cost of a spot commercial comprises a time charge only. Television advertising rates are divided into four time segments. The most expensive period is the A segment which covers prime time, from 1900 to 2300 hours. Second is the "Special B" segment, covering the time periods just before and just after prime time, plus 1200 to 1400. The charge for advertising during this segment is usually about 70 percent of the A segment charge. Third is the B segment, which covers the rest of the daytime viewing, costing approxi-

TABLE 10.2. Magazine advertising rates: 1988
(Yen per back cover advertisement).

General – weekly	
Shukan Asahi	2,300,000
Shukan Yomiuri	1,500,000
Mainichi Graph	570,000
Women's – Monthly	
Ie no Hikari	3,600,000
Fine	1,500,000
Cooking	
Eiyo to Ryori	1,050,000
NHK Kyo no Ryori	3,400,000

Source: Dentsu Japan (1989), pp. 307-308.

mately 40 percent of A time. Finally, the C segment covers early mornings and late nights, costing about 30 percent of the A time rate. Table 10.3 lists some television advertising rates for the A time period.

The fourth main medium for advertising is provided by radio. In 1987 Y172.7 billion was spent on radio advertising (Dentsu Japan, 1989). Table 10.4 gives examples of radio advertising rates for 30-minute program sponsorships and for 20-second spot advertisements.

A beef exporter who wished to advertise in Japan would probably have to operate through a Japanese advertising agency. There are many agencies but the industry is dominated by Dentsu Japan, which accounts for about 25 percent of all advertising expenditures in Japan. An advertising agency will ensure that a product is advertised in a "Japanese" way. Western slogans that have been used to promote a product in an exporter's domestic market are often unsuitable for the Japanese market. The Japanese language is complex: There are four ways of writing Japanese, each with different

TABLE 10.3. Television advertising rates: 1988
(Yen – during A time periods).

Station (VHF)*	Time-30 Minutes	Spot-20 Seconds
Tokyo-NTV	1,500,000	750,000
Tokyo-TBS	1,600,000	750,000
Hokkaido-HBC	1,200,000	420,000
Kinki-ABC	1,300,000	600,000
Okinawa-OTV	700,000	200,000
Kyushu-OBS	800,000	200,000

*Rates for UHF television stations are similar.
Source: Dentsu Japan (1989), p. 342.

but subtle nuances that would create different product images if used in an advertisement. It is therefore unlikely that an advertising campaign in Japan could be prepared without the services of a Japanese advertising agency. Obviously, this adds to the costs of marketing the product since neither form nor substance can be transferred from the exporter's domestic market to Japan.

A comprehensive and long-term promotional campaign could become an essential ingredient to the successful marketing of imported beef in Japan. Exporters need to create an awareness of their beef, differentiating it from that of competitors. This may take the form of advertising in one or more of the four major media, advertising on outdoor signs or on public transit, participating in food exhibitions, or organizing cooking demonstrations and producing beef cookbooks. The services of a Japanese advertising agency are probably necessary.

Promotional activities, though costly, are probably essential. As De Mente (1988) explained:

> The Japanese are far more advertising-oriented than are consumers in many other countries. It is therefore necessary for a company that wants to maintain or increase its market share to advertise continuously and more heavily than is usually the case elsewhere–in as many media as possible. (p. 198)

TABLE 10.4. Radio advertising rates: 1988 (Yen).

Station	Time-30 Minutes	Spot-20 Seconds
Tokyo-NBS	500,000	88,000
Tokyo-NSB	420,000	42,000
Hokkaido-HBC	290,000	50,000
Kinki-JOCR	250,000	36,000
Okinawa-RBC	120,000	15,000
Kyushu-OBS	100,000	15,000

Source: Dentsu Japan (1989), p. 342.

The final price realized by the exporter is what is left over after all of the production, processing, importing, distribution, and promotion costs have been deducted, and an allowance has been made for the profits of all those in the vertical coordination chain.

SUMMARY

The complex distribution system in Japan leads to considerable differences in the import and retail price. As a result, it is necessary for firms wishing to export to Japan to identify the point where its product should enter the distribution system so as to minimize those costs. Market intelligence and market promotion are extremely well developed in Japan. Firms wishing to create an image for their product, or to increase the market for their beef, would probably be wise to hire the services of a Japanese marketing firm.

Chapter 11

Wholesale and Retail Prices

There are three features of Japanese home consumption of beef that set the market apart from those in other countries: (1) high overall prices relative to those in other countries, (2) long periods of price stability, and (3) a wide price range for beef of different grades, cuts, and origins. While at first glance these characteristics may appear contradictory, they can be easily explained in the context of the highly complex and heterogeneous wholesale and retail markets.

HIGH LEVEL OF PRICES

Comparisons of retail beef prices among nations must be made with a high degree of caution. To have a truly useful comparison, one must compare products from similar quality carcasses, that are cut to the same specifications and are presented to the consumer with the same degree of final finishing by the retail butcher. Given the great diversity of tastes for beef internationally, and the wide variety of butchering procedures arising from different culinary requirements, providing comparisons that reflect true differences in value is extremely difficult. As most exporters actually enter the market at the carcass or boxed beef stage of the marketing chain, international comparisons of input prices–producer prices for carcass beef–are a logical place to begin. Unless it can be assumed that the Japanese processing or distribution sector is more efficient than those in other countries, input price differentials should be an indication of the minimum price differential one might expect throughout the marketing chain. As the Japanese distribution system is known to be less streamlined than those in other developed countries, the differential will likely be even greater.

An international comparison of producer prices for beef is presented in Table 11.1. Japanese producers clearly receive very high prices for their product, almost three times that which U.S. processors have to pay for their raw materials, and over seven times the Australian price for raw materials.

A comparison of Japanese wholesale prices with French wholesale prices in 1982 showed that Japanese prices ranged from Y2,400/kg to Y10,000/kg (Namiki, 1990). Observations at a morning meat market in Paris reported that prices for specific cuts in the French market ranged from Y2,200/kg to Y2,700/kg. Prices of similar cuts in Japan ranged from Y3,000/kg to Y18,000/kg (Ogura, 1983).

Retail beef prices in Japan have been higher than retail prices for other meat products (Table 11.2). This price differential is the result of the limited domestic supply of beef, the restrictions on imports, and the price stabilization policy of the Japanese government, as well as growing demand for beef in Japan, placing upward pressure on prices.

The lack of official retail price data at a sufficiently disaggregated level makes it difficult to trace any changes in retail prices of imported beef. Nakase (1990) reported that retail prices of frozen shoulder cuts of imported beef were declining. From a yearly average of Y145/100g in JFY 1986, they declined to Y128 in 1987 and to Y118 in 1988. Retail prices fluctuated between Y111/100g and Y121/100g throughout 1989, averaging Y118 over the year. In general, since the BMAA, imported beef prices have been falling, domestic dairy beef prices have leveled off, and Wagyu beef prices have risen (Namiki, 1990).

Retailers may be unwilling to further reduce imported beef prices (Mori and Lin, 1990). The demand for imported beef tends to be unresponsive to changes in its price.[1] Hence, retailers may not have a strong incentive to drop imported beef prices substantially in the near future since their total revenue would fall. To be more competitive with domestically produced beef, imports may have to compete on the basis of quality and product differentiation. The lack of retail

TABLE 11.1. Producer prices for beef 1991
($/100 kg carcass weight).

Japan	814.93
Switzerland	534.24
Sweden	316.16
U.S.	278.60
United Kingdom	239.99
Canada	213.91
New Zealand	115.82
Australia	110.60

Source: OECD (1992a).

price data prevents a more explicit forecast of future retail prices for beef in Japan.

LONG-RUN STABILITY IN PRICES

Since the inception of the beef industry in Japan, beef producers have been well insulated from world market forces. The Japanese government has operated a price stabilization program which has kept carcass prices, and thus farm-level prices, within a relatively narrow band. As can be seen in Table 11.3, the lower and upper price bands have moved very little over a 15-year period. Every year the price limits are set by the government, following negotiations with the representatives of livestock producers.

In addition to the stabilization policy for beef prices, import controls administered by the LIPC prior to 1991 dampened any pressures from the international market that might have initiated beef price changes in Japan. The lack of competitiveness in the transportation, handling, processing, and marketing sectors has meant that relatively stable carcass prices were reflected in relatively stable retail prices. In Table 11.4, it can be seen that retail beef prices have moved upward in a very leisurely fashion and in concert with small changes in wholesale carcass and live cattle prices.

TABLE 11.2. Retail prices for various meat products (Tokyo) ($/100g).

	Beef*	Pork	Poultry**
1986	2.78	1.18	0.88
1987	2.80	1.16	0.83
1988	2.82	1.16	0.83
1989	2.54	1.03	0.73
1990	2.84	1.13	0.79
1991	3.12	1.24	0.87

*Medium Grade
**Boneless Broilers
Source: Japan MAFF (1992)

For a variety of reasons, this relatively stable price regime for beef can be expected to be maintained after the removal of import quotas and the further opening of the Japanese beef market. First, the Japanese government is unlikely to abandon its price stabilization program for cattle producers in the face of strong agrarian political pressure. Indeed, since signing the BMAA in 1988, the Japanese government has acted to strengthen domestic protection for Japanese beef producers. Second, the relatively high c.i.f. tariff of 50 percent on imported beef products, combined with the high transportation costs associated with shipping foreign beef to Japan, limits the exposure of the Japanese beef industry to international market forces. Third, the highly regulated processing, distribution, and retailing sectors leave little room for the hope that a rapid expansion of beef imports would force Japanese beef prices down and thus spur increased consumer demand for beef products. Of course, multilateral negotiations may eventually lead to not only better access but also less shielding of Japanese beef prices from international market forces. Certainly the United States, Australia, and other beef-exporting countries can be expected to apply heavy pressure on the Japanese government to reduce the protection afforded their domestic beef industry.

TABLE 11.3. Stabilization price ranges for beef carcasses (Y/kg).

Japanese Fiscal Year	Wagyu Steer Medium Grade		Dairy Steer Medium Grade	
	Lower Price	Upper Price	Lower Price	Upper Price
1975	1,143	1,518	930	1,236
1976	1,240	1,647	1,009	1,341
1977	1,303	1,730	1,061	1,408
1978	1,303	1,730	1,061	1,408
1979	1,303	1,730	1,061	1,408
1980	1,357	1,763	1,105	1,435
1981	1,399	1,817	1,118	1,452
1982	1,400	1,820	1,120	1,455
1983	1,400	1,820	1,120	1,455
1984	1,400	1,820	1,120	1,455
1985	1,400	1,820	1,120	1,455
1986	1,400	1,820	1,090	1,420
1987	1,370	1,780	1,020	1,325
1988*	—	—	995	1,295
1989*	—	—	995	1,295
1990*	—	—	985	1,285

*Steer Grade B-2, B-3 (following change in grade classification, 1988).
Source: Japan MAFF (1992).

WIDE RANGE OF PRICES

One of the most striking aspects of the Japanese beef market is the wide range in retail prices. Kerr and Klein (1989) observed that:

It is common to see beef prices in a supermarket which range from $10 per kilogram to $200 per kilogram. (p. 40)

It has been claimed that Wagyu beef prices range from $66/kg to $330/kg, yet U.S. strip loin—one of the higher quality imported beef

TABLE 11.4. Cattle and beef prices at different stages (Y/kg).

Year	Live Cattle*	Wholesale Carcass**	Retail Beef**
1980	797	1.329	3.390
1981	723	1,245	3,360
1982	736	1,304	3,420
1983	739	1,298	3,510
1984	735	1,284	3,570
1985	747	1,318	3,510
1986	778	1,339	3,530
1987	756	1,289	3,550
1988	740	1,232	3,550
1989	791	1,250	3,850
1990	723	1,256	3,830
1991	N.A.	1,210	3,910

*Dairy steer.
**Medium grade dairy steer until 1988; B-2, B-3 grade beginning in 1988.
***Medium grade.
Source: Japan MAFF (1992).

cuts–was observed to be priced at approximately $15/kg in a department store in downtown Tokyo (Mori, 1990).

Retail beef prices in other countries do not diverge by these large amounts. Beef in Japan is an extremely differentiated product with a wide range of prices and qualities available. The range of retail beef prices arises largely because of the perceived quality differences between Wagyu, dairy, and imported beef, between grain-fed and grass-fed beef, and also between chilled and frozen beef. Most domestic beef is identified by place of origin and strong efforts are made to promote these "differentiated" products. Thus, a particular cut of grade A4 Wagyu beef produced in the Kobe region of Japan may well sell at a far different price than would the same cut of A4 Wagyu beef produced in the Niagata region. Often, even dairy beef is identified by place of origin and promoted on the basis of "superior" lineage of parents, feeding methods, etc.

The diversity in prices is also evident at the wholesale level. There are 15 different Japanese beef grades. Within these grades, a distinction is made between Wagyu heifers and Wagyu steers and between dairy heifers and dairy steers. Consequently, there may be at least 40 different wholesale classifications for domestically produced beef in the Japanese market.[2] When chilled beef and frozen beef from a number of countries are added, the entire market takes on the appearance of a continuum from very low to very high quality and prices.

Of course, the high prices observed in the high-quality portion of the Japanese beef market is one of the major attractions of the Japanese market for exporters. Hence, if an exporter could produce a beef product that falls into the higher end of the Japanese beef quality spectrum and could minimize production, processing, transportation, distribution, and marketing costs, then the Japanese beef market could prove extremely lucrative. However, exporters face high costs for producing heavily marbled beef, processing and packaging it to Japanese specifications, transportation, tariff charges, and distribution and marketing.

Western visitors to the meat counters of Japanese supermarkets are invariably amazed at the range of prices and types of meats on display. In picture plate 1, superior-grade Kobe beef roast steaks (Wagyu) are on sale in a Tokyo department store for Y3,000 for 100 grams. This is Y30,000 ($240) per kilogram. As can be seen from the picture plate, most Westerners would not consider these steaks to contain much beef; they look to be mostly fat. In supermarkets outside Tokyo, it is not uncommon to see similar quality Kobe beef priced at Y4,000 or Y4,500 per 100 grams.

Picture plate 2 illustrates some lower qualities of Kobe beef. The top three items, priced at Y600, Y800, and Y1,000 per 100 grams, respectively, are beef prepared for Shabu Shabu. Note the thin slices used in the Shabu Shabu dish. The three bins are graded as regular, medium, and superior, in line with their prices. The regular and

PICTURE PLATE 1. Superior-grade Kobe beef roast steaks (Wagyu).

PICTURE PLATE 2. Lower qualities of Kobe beef.

medium qualities are thigh meat; the superior quality, for Y1,000, has been cut from a shoulder roast.

In the middle panel of picture plate 2, stew meat is on offer. Regular-grade Kobe stew meat is priced at Y400 per 100 grams and curried Kobe stew meat is priced at Y500 per 100 grams. The display on the right side of picture plate 2 (middle row) is American sirloin steak priced at Y600 for 100 grams ($4.80/kg).

Picture plate 3 is an Ochugen (gift) corner. Two times during the year–mid-summer and year-end–many Japanese present gifts of meat to business associates, employers, and special friends. This picture was taken in August, so these displays were for mid-summer gift giving. In the top right-hand corner there are six filets, with a total weight of 900 grams, selling for Y25,000 ($200). Next to this package is a package containing three filets, with a total weight of 450 grams, plus two small roasts, with a total weight of 400 grams, selling for Y20,000 ($160). Third from the right is a package containing five sirloin steaks, each weighing 200 grams, selling for a price of Y15,000 ($120).

The three picture plates represent only a small sample of the wide

PICTURE PLATE 3. An Ochugen (gift) corner.

variety of beef cuts, qualities, and prices that can be observed on any visit to a meat counter in Japan. Different qualities of hamburger often sell for prices between Y200 and Y400 per 100 grams ($16.00 - $32.00 per kg). This wide variety of cuts, qualities, and prices provides potential exporters of beef with many opportunities to fill different market niches.

In the past, it has been difficult for exporters to keep track of their beef once it entered the Japanese beef distribution system. This has made it difficult for exporters to know the final market for their products and, in particular, to know the final retail price of their beef. In fact, reliable data about retail prices of various beef cuts has not been readily available (Mori and Lin, 1990; Namiki, 1990). The only official source of retail prices has been the "Retail Price Survey" published by the General Affairs Agency (Namiki, 1990). This survey, however, reports "average" prices for domestic and imported beef. The data fail to differentiate between Wagyu and dairy beef and between grain-fed and grass-fed beef, thereby limiting the data's usefulness. Improvements in the availability and content of retail price data for the Japanese beef market may become an important requirement of successful export marketing in the future.

SUMMARY

By any standards, some retail beef prices in Japan are extremely high. At the same time, some of the beef products that would be considered most desirable in the markets of exporting countries can be found at a modest price in Japanese supermarkets. The key for foreign exporters remains finding that niche in the quality continuum for which a product can be profitably tailored. Information is poor and being able to ensure that the product actually reaches the niche is likely to prove difficult. If a profitable product can be produced and marketed, the exporter is likely to face a fairly stable market price over considerable periods of time.

Chapter 12

Beef Consumed Away from Home

In the previous two chapters, the discussion of beef consumption in Japan focused on the quality and pricing of beef to be consumed in the home. Implicit in this discussion has been the assumption that the beef would also be cooked in the home. There is, however, a significant and growing market for beef as an input to meals that are professionally prepared and consumed outside the home. This market has become known in the meat trade as the hotels, restaurants, and institutional (HRI) market. The strict division between the home market and the HRI market is somewhat blurred because of the home delivery and takeout trade. In these cases the food is professionally prepared in a non-home setting and consumed in the home. The beef market in Japan has become a virtual continuum as fully prepared meals are also sold in retail outlets as oven-ready. These types of products have also experienced considerable growth as a result of the widespread use of microwave ovens. These meals simply require reheating rather than cooking in the true sense of the word. Hence, there is a very fine line between a pizza purchased at a restaurant, picked up at a takeout window, and reheated upon arrival at home, and a prepackaged pizza bought in a retail food outlet and heated in the microwave at home.

In this chapter, the market is arbitrarily divided. The pizza purchased from a restaurant and taken home to be eaten is considered to be part of the HRI market, whereas the prepackaged pizza bought in a retail food outlet is considered to be part of the home consumption market. This seems a logical point of division since, regardless of whether the food is consumed in the home or away from the home, if the meal is prepared in a restaurant or similar institution it will draw upon the same sources of meat supply and have the same quality specifications.

HOTELS, RESTAURANTS, AND INSTITUTIONS

The HRI trade has, to some extent, been arbitrarily classified into its three components. Often, the requirements of hotels and restaurants are nearly identical in the high-class segment of the market because hotels tend to have relatively high-quality restaurants. In fact, restaurants physically located in hotels may be leased or contracted to firms that act as de facto private restaurants. A more appropriate division might be between high-class restaurants (located either within or outside of hotels) and fast-food restaurants since the beef required by the fast-food industry differs considerably from that used in high-class restaurants.

In the past, hotel restaurants were places where people in the community went to eat on special occasions; in recent years, however, hotel restaurants have faced considerable competition. As incomes have risen, more eating establishments could be supported in a community. This means that the hotel's room clientele are no longer a captive market for the hotel's restaurant. While hotel guests may have their breakfast in the hotel, lunch and dinner can easily be consumed elsewhere.

To a large extent, fast-food restaurants are a phenomenon of the post-war era. They are a result of three major forces. First, assembly-line industrial production techniques have been adapted to the food industry. This has allowed standardization and economies of scale in the production and/or purchasing of inputs, the use of unskilled labor for food preparation, and short turnaround times between food order placement and food consumption. Second, as incomes have risen, people have begun to place a relatively higher value on the time traditionally spent in food preparation. Hence, they have been willing to pay for the convenience of having that time free for other leisure activities. Third, the increased rate at which women participate in the labor force has meant that women have less time available for traditional activities such as cooking. As a result, hard-pressed single parents and families in which both parents work are more willing to opt for the convenience of restaurant food. These factors have been augmented by greater employee travel and the employee working in a number of locations rather than one. In other words, as the service sector has expanded relative to manufacturing, workers have been presented with greater opportunities for restaurant food consumption.

Manufacturing tends to be sedentary, whereas services tend toward personal visits to the place of business. Computer repair technicians, photocopy service personnel, consultants, and those contracted for the maintenance of a building's shrubbery–all move from place to place during the day. As a result, lunch and perhaps dinner are taken in fast-food restaurants rather than in the traditional lunchbox prepared at home and taken to the place of work for consumption. Often, meals taken away from the business headquarters are considered to be business expenses and are paid for by the employer.

The third segment of the market, institutions, is a catch-all term for a large number of venues in which food is not consumed in the home. At one end of the scale are military establishments, prisons, hospitals, and airlines. The consumer here has little choice about the place where food is consumed or the cuisine. Often they are institutions in which no direct payment is made for food or where food is considered a portion of the customer's daily cost. These institutions tend to be concerned primarily with providing a set level of nutrition at a minimum cost. Other institutions allow for choice by the consumer, but in effect they provide a subsidy sufficient to make alternatives unattractive except in the case of special occasions. School cafeterias, university food services, and company canteens fall into this category. The consumer may incur no direct cost, but most often these institutions require some direct customer payment for food. They tend to minimize cost in order to reduce the subsidy required, but they are somewhat constrained by their customers having the opportunity to exercise free choice. Finally, there are institutions that are not subsidized and operate in largely competitive environments–hotels and halls catering to conventions, lunch meetings, and weddings. While the market environment within which such institutions operate can vary considerably, they are all characterized by set or limited choice menus, large volumes, and prearranged group feedings.

The evolution of the HRI trade has followed a similar pattern in most developed countries. Rising incomes affect this sector's rate of growth. Japan's spectacular income increases in recent years have meant a rapid expansion of the HRI sector. The rate of increase in consumption has lagged somewhat behind the rise in income. In part, this reflects the degree to which Japanese society has been able to adjust to the new, higher level of income.

The strictly institutional portion of the Japanese HRI market has been fairly stable over time. However, as with other developed countries that have experienced rapid growth in consumer incomes, Japanese consumers have progressively taken a greater proportion of their meals outside the home-cooked and home-consumed setting. Starting in the 1960s, the Japanese restaurant industry has experienced considerable and sustained growth. As Japanese incomes have grown at a rapid rate, they are now one of the wealthiest societies in the world. Average annual household income exceeded US$50,000 in 1990. Approximately 85 percent of this income can be considered discretionary–that is, after taxes and other social welfare deductions. The food service and leisure industries capture a large portion of the increase in food expenditure that results from income growth. This is because the basic food, shelter, and consumer durables markets can, for the most part, be considered mature industries.

All of this translates into a large percentage of meals eaten outside the home. Between the mid-1970s and the mid-1980s, total sales in the out-of-home preparation market nearly doubled, from US$100 billion to US$185 billion. Approximately 35 percent of all expenditures on food take place in the HRI sector. Of course, this tends to overestimate the proportion of food consumed outside the home-prepared and home-consumed segment of the market since these expenditures include all of the payments for service and ambience that are included in the price of HRI food. The percentage of extra income spent on eating out is 50 percent higher than that for food purchased for preparation and consumption in the home.

Part of the reason for the rapid rise in restaurant sales can be explained by the Japanese approach to entertaining guests or friends. Japanese houses tend to be very small and, hence, not particularly suited for entertaining. So, rather than entertaining in the home, the Japanese typically entertain in the facilities of their favorite restaurant. The restaurant industry provides a large variety of establishments from which prospective customers can choose.

By Western standards, most Japanese restaurants are small. Seating capacity is very limited, between 5-15 people on average. Their staffs also tend to be very small, usually three or four people.

The food service industry, however, has remained relatively labor-intensive. As a result, less than 40 percent of the value of sales can be accounted for by the cost of food used in restaurant meals.

The market is divided into restaurants serving either Japanese or Western-style cuisine. Traditional Japanese restaurants use more fish on their menus while the Western-style restaurants use a greater proportion of meat–often constituting as much as 40 percent of the value of the food purchased as inputs to the meals. The institutional segment of the HRI market utilizes larger proportions of rice and other processed foods.

Approximately 10 percent of all food imports are used by the HRI sector. Western-style family food restaurants and fast-food chains, however, use a much higher proportion of imported foods as inputs–as high as 70 percent in some cases. These factors have significant implications for the quantity and quality of beef imports.

QUALITY AND SOURCES OF SUPPLY

The restaurant industry in Japan can be partitioned into those that cater to traditional Japanese tastes and those that cater to Japanese interpretations of Western cuisine. High-quality Japanese restaurants, Shokudo restaurants, sushi bars, noodle shops, and ryotei cater to Japanese tastes. Those serving Western cuisine can be divided into high-class Western ethnic restaurants, Chinese restaurants, family restaurants, and fast-food establishments (Department of External Affairs, 1988). In between are pubs or bars that serve limited menus of both Japanese and Western meals. It is the Western portion of the market that has exhibited the greatest growth in recent decades.

High-class Western restaurants in Japan tend to mirror those in any part of the world. They tend to specialize in ethnic cuisine–French, Italian, American, German, Spanish, or their regional derivatives–Parisian, Tuscan, Californian, Bavarian, Basque. Such establishments tend to be expensive and will purchase high-quality inputs, including beef. Their volume is limited and they are usually not a part of a chain, meaning they don't have a contract system of purchasing. Hence, they purchase specialty items or portion cuts. Their suppliers usually deal in imported chilled product, usually from North America. If Western exporters can establish distribution channels for sub-primal or portion-ready cuts, this market may expand considerably. The marbling characteristics of North American beef are well suited to this type of cuisine.

In addition, these establishments often require some specialty products, particularly veal. Continued growth should put additional pressure on the Japanese dairy herd to slaughter surplus bull calves as veal rather than retain them for feeding and slaughter as mature animals. While not a fast growth segment of the market, sales have been growing at approximately 10 percent per year. If Japanese incomes continue to grow as they have in the past, a larger proportion of consumers will be able to experiment and identify their individual tastes for the various ethnic foods. Since these ethnic foods are mostly Western, and therefore often based on large beef portions (by Japanese standards), they will represent a growth market for both foreign and domestic beef in Japan.

While not Western, Chinese restaurants do represent an alternative to Japanese cuisine. Many of these establishments can be characterized as the bottom end of the consumer market. Most meals in these restaurants are noodle- or soup-based. If meat is used at all, it tends to be very low priced, with chicken predominating. There may be some growth for beef in this market, but it is likely to represent the scraps and other by-products of the butchering process as the total market for beef expands and greater quantities become available.

Probably the fastest growing segment of the Japanese restaurant market is what can be termed Western-style family restaurants. Recently, growth rates have been as much as 20 percent per year (Department of External Affairs, 1988). These establishments are usually patterned after franchises of North American chains. The menus are limited and they rely on relatively high volumes. As a result, they can use staffs with limited in-house training. While many of these establishments are found in the traditional locations for restaurants in Japan–near train stations and along busy pedestrian thoroughfares–they have been particularly aggressive about catering to the driving public. This has meant that they have, to some extent, been able to escape the high-rent areas that have traditionally forced restaurants in Japan to have very restricted space. As a result, they are generally able to provide purpose-built restaurants that are airy, modern, spacious with adequate parking and reasonably priced food because they pay relatively low rents.

The menus usually specialize in Western dishes, meaning that they purchase considerable quantities of hamburger meat and steak. The source for the hamburger can be either domestic or frozen imported beef. Price is a major determinant in the purchase decision. The steak dishes in these restaurants are able to use, as in North America to a considerable extent, either frozen grass-fed product or lower-quality beef from domestic dairy animals. If the beef price falls in the wake of liberalization, then a greater proportion of beef dishes can be expected on the menus of these establishments. This should lead to continued rates of growth, even if the growth in this sector cannot be sustained at its current high levels. These restaurants represent over half the Western-style restaurants in Japan and outstrip fast-food establishments in both sales and growth. While there has been considerable attention paid to the success of U.S.-based fast-food chains (such as McDonalds) in the Japanese market, it is in the family restaurants where the real taste for Western food has been manifest. It is also the sector in which "Japanese-style Western food" has been evolving.

Fast-food chains, either franchise operations of foreign chains or indigenous Japanese copies of such chains, have had considerable success in the Japanese market over the last two decades. It has been the hamburger chains that have been the most important influence on the acceptance of Western food in Japan. Many chains have in excess of 100 outlets. In the 1970s and 1980s these chains exhibited extremely rapid rates of growth in sales, often in excess of 50 percent per year (Department of External Affairs, 1988). With the eventual "packing" of this market segment as the industry approaches maturity, the industry's growth rate has declined.

While these fast-food outlets are often considered to be at the bottom of both the price and quality continua for food establishments in North America and Europe, they are perceived as holding a position somewhere further up those continua in Japan. The reasons for this are a combination of the following: the prestige attributed to things Western in Japan, the use of different marketing strategies than those used in the West, and the existence of an even lower-quality traditional Japanese segment of the industry. The lowest segment of the market is held by noodle shops, which are patronized in a very similar way to fast-food establishments in the West.

The Japanese have a love-hate attitude towards things Western. While the Japanese are very chauvinistic about the quality and importance of things Japanese, and as a result denigrate products from elsewhere, they also perceive Western products as modern or trendsetting. Hence, while not being willing to significantly reduce their commitment to traditional Japanese cuisine, dietary concessions are allowed so that an image of modernity may be projected. Fast-food restaurants serving Western cuisine, including wholly Japanese chains following the Western formula, have benefitted from this need. Further, fast food has the advantage of representing an easy transition to a Westernization of eating habits. Since Japan has a chopstick-based cuisine and table etiquette, many Japanese find the full transition to a Western cuisine knife-and-fork-based table etiquette very intimidating. Fast food, with its "hands only" table etiquette, represents a face-saving alternative route to the appearance of Westernization.

The marketing strategies of Western fast-food chains have been aimed at reinforcing an up-tempo and modern image. The first outlets were in the Ginza and other prominent places in Tokyo where the well-to-do modern corporate Japanese congregate. Many Japanese corporate partners of Western fast-food chains have been associated with high-quality products in other areas–McDonalds with the prestigious Mitsukoshi Department Store, Kentucky Fried Chicken with Mitsubishi, Wimpy International with Toshoku. As a result, these corporations have been able to capitalize on their image to attract customers in the higher income groups.

Much of this business is franchised. Between 1975 and 1988 the number of food service chains increased from 77 to 294. The total number of stores exceeded 36,000 in 1988, a sixfold increase over the period (Department of External Affairs, 1988). Five out of every six stores was a franchise. By 1986 there were over 2,000 hamburger chain outlets alone.

While this segment of the restaurant trade is growing at a much slower pace than in the past, there is one factor that may significantly increase growth for beef in the wake of import liberalization. In the period of regulated imports, quotas were valuable and distributed to a select set of end-users. The quotas were too valuable to use for imports of large quantities of hamburger meat since the markups that arose from the import quota system were much greater for higher priced cuts. As a result, imported hamburger meat was both very expensive and scarce. This had a positive effect on the price of that domestically produced beef which was suitable for hamburger. Pork could be imported in unrestricted quantities and was subject to a much lower tariff than beef. As a result, much of the hamburger sold in fast-food chains was not pure beef but rather a blend of pork and beef. With the removal of import quotas, it is possible to import large quantities of beef of suitable quality for hamburger. If the relative price of hamburger falls, there is likely to be substitution in favor of a pure beef patty. This could provide considerable growth in beef sales over and above what could be expected from industry growth alone.

Since neither the dairy cow herd nor the Japanese beef cow herd is likely to significantly expand in the future, domestic sources of hamburger-quality beef are likely to remain fairly constant. Hence, imports will be needed to fill the market growth. The hamburger industry is, however, price–rather than quality-driven and, hence, the major beneficiaries of the fast-food market growth will likely be Australian and New Zealand grass-fed beef. Since North America is a long-term net importer of hamburger beef, neither the U.S. nor Canada will be able to profitably exploit the hamburger market.

Competing in the same market as Westernized fast-food restaurants are Japanese restaurants commonly known as Shokudo. These restaurants are basic eating establishments with little atmosphere. The food is Japanese–miso soup, rice, pickled vegetables–with fish dishes predominating, but hamburger, steak, and beef curry are often available.

Shokudo tend to be small, with two or three employees. This sector has not benefited much from the rise in Japanese incomes and the industry has been growing very slowly. Hence, this segment of the market is not likely to provide a significant growth market in the future.

Below the Shokudo in the Japanese price-quality continuum is the noodle shop. The major protein source in these restaurants is soya, which is included as either a soup base or a dip for the noodles. Some of the soup bases are beef stock and there is a stable market for some low value beef cuts. This market has been shrinking very slowly in recent years. As incomes increase, noodle shops' traditional customers have been increasing their patronage of Western fast-food chains. Equally unproductive as a market for beef are Sushi bars. Fish predominates as the protein staple and, of course, raw fish is the product upon which the concept is based. However, beef sashimi (raw beef) has been made available in some very expensive sushi restaurants in recent years.

High-quality Japanese restaurants are quite formal, with waitresses in traditional dress. While much of this cuisine is fish or seafood based, it is also the market niche for the heavily marbled Wagyu beef. These restaurants have been experiencing a strong growth in sales. These establishments are seldom part of chains, so the quantities of beef required are very small for each establishment. As a result, the cost of distribution to this sector is very high. There is little doubt that these restaurants, near the high end of the price-quality continuum, will continue to rely exclusively on the domestic Wagyu sector for its beef inputs. This is particularly true for those restaurants that specialize in the small shimofuri beef market. Restaurants considered to be at the bottom or even the middle range of this segment may provide a market for long-fed North American beef if it can be priced below domestic Wagyu. The product will have to be available in a chilled form to retain the highest cooking qualities. If off-shore well-marbled beef can be produced and delivered at a lower price, then some expansion into this sector may be possible to take advantage of this quality of beef.

At the height of the price-quality continuum are the establishments known as Ryotei. These are extremely expensive, an evening often costs more than US$500 per customer. As a result, these establishments are patronized almost exclusively by those on expense accounts. Given the prices charged, the price of food inputs is of little concern relative to high quality. This segment of the market has been shrinking, however, as its clientele ages. Hence, it seems unlikely that it will be a significant growth area in the future.

It should be clear that growth in the restaurant industry is concentrated in the Westernized portion of the market. To the extent that rising incomes are translated into increased food expenditures, including beef, it would seem that the extra Japanese food expenditures are made to improve the quality of food prepared in the home while extra restaurant expenditures are concentrated in the market for Western cuisine. This is because Japanese wives are not familiar or comfortable with the preparation of Western cuisine and Japanese kitchens are not equipped to produce Western meals. Hence, it is easier to broaden one's diet by going to Western restaurants. Since Western food is more heavily based on beef, this is likely to lead to expanding markets for beef in this sector.

SUMMARY

It is clear that the major area for growth of beef sales in the Japanese HRI trade is likely to be Western-style restaurants. Given the diversity of restaurants, the type of beef required varies greatly in quality. The only areas in which significant growth is unlikely are in the lowest and highest quality restaurants. All other areas can expect good growth. Marketing success will be based on establishing close ties with the middlemen who do the provisioning for the restaurant industry. Price is likely to be the most important factor for securing success.

PART IV: PROSPECTS

What are the prospects for those who might wish to consider investing in supplying the Japanese beef market? Clearly, it is a market that is both growing and changing. While growth suggests opportunities, change implies risks. As this book outlines, there are a myriad of dynamic forces at work in the Japanese beef market, and the economy as a whole, that interact to shape beef consumption patterns. Can any predictions be made about this complex market upon which to base business decisions? In in this section an attempt is made to answer these questions. Chapter 13 looks at what can be learned from professional forecasts, while Chapter 14 outlines the major marketing issues.

Chapter 13

Forecasts of Market Growth

While it may have been possible to identify the major factors that affect the Japanese consumption of beef, the question of how the Japanese market for beef will develop over the near term remains unanswered. This has been deliberate. The discussion of the Japanese beef market in previous chapters has provided some insights, but concrete forecasts have not been made. This is, in part, a function of the nature of the forecasting procedure itself. Forecasting market developments can be confidently undertaken only under a fairly limited set of circumstances. The Japanese market for beef meets few, if any, of these conditions and, hence, existing forecasts must be scrutinized carefully, keeping in mind the assumptions made in producing the prediction.

THE LIMITATIONS OF FORECASTING

While a large number of methods can be used to forecast market developments, the two methods most often used for the Japanese market (after import liberalization) are armchair speculations and econometric modeling. While out of fashion with a large segment of academics who are familiar with advanced statistical techniques, armchair speculations remain the major means by which predictions are made by marketers and business people. As with the more technical forecasts, armchair speculations vary in quality to a considerable degree, depending upon the abilities of the individuals making the forecast and the resources they have available. Note that the word *quality* is used to describe forecasts and not the word *accuracy*. The quality of a forecast may have little to do with its accuracy because accuracy can be measured only after the fact–that is, by comparing forecasted values with what actually occurred. Any market can be significantly affected by forces that cannot be predicted before they happen and, hence, cannot be included in the reasoned assessment of a market's future development. In short, a poor forecast may be accurate by accident, while a good forecast can be far off the mark due to unforeseen circumstances.

A good armchair prediction relies primarily on the "wisdom" of the individual(s) making the forecast. While there may be a number of definitions of wisdom, the one visualized here is of an individual who has experience with the market, good analytical skills, and the knowledge of how to acquire the requisite data. Armchair speculation is probably best undertaken when markets are simple and relatively stable. Under these circumstances, experience can provide sufficient insight to determine the significant pieces of information required for a reasoned prediction. Typically, such expertise can be acquired over time. It is most easily acquired in markets for products that can be considered homogeneous; that is, of a single quality. Further, the underlying structure, both technological and regulatory, must be stable. The economic forces at work will then become clear to the forecaster over time.

The Japanese beef market meets neither of these criteria. If this book suggests nothing else, it is that marketing beef in Japan is complex and not simple. The product is far from homogeneous, and beef consumption in Japan takes place over a very wide spectrum of quality. As a result, the information requirements of the armchair forecasters exceed both the time required for ingestion and the ability needed to accurately assess it. Further, the deregulation of the importing system means that the significant factors affecting the post-deregulation market are opaque. Until the market approaches equilibrium, it will be difficult, if not impossible, to determine the direction the market is taking.

This does not mean that there cannot be a general consensus regarding the significance of some forces. There is little argument that as incomes rise over time, beef consumption will increase. This observation, however, is somewhat trivial because the important question is how will the various segments of the beef quality continuum evolve. Some market segments will clearly expand faster than others, and market shares will shift among segments. These multidimensional interactions are very difficult to assess before the fact.

It might be possible to acquire expertise in one market segment if the technological and regulatory regimes remain stable. Since rapid technological change is not manifest in the beef industry, it does not present a particular problem. The regulatory regime, on the other hand, is changing rapidly.

Removal of quantitative restrictions on beef imports and raising of the tariff has led to speculations regarding future Japanese retail prices. These speculations range from a decline to world price level plus the tariff and competitive handling, butchering, and retail markups to no decline in retail price due to the oligopolistic nature of the distribution system. In the latter case, it is believed that the Japanese processing and distribution system has sufficient market power to simply capture the proportion of the markup that had previously accrued to the LIPC. Probably the truth lies somewhere in between the two extremes.

Since this question cannot be answered with any degree of confidence, it means that any analysis of the effects of price declines cannot be based on information from historic interactions of prices in ranges lower than have prevailed in the past. Since grades of beef along the quality continuum are substitutable at the extremes of their ranges, it is not possible to specify quantity changes in the absence of information on price changes all along the continuum.

It *is* possible to use econometric modeling techniques to capture the interactive nature of the changes in a number of prices. Econometric techniques, however, are not particularly helpful for analyzing the continuum directly. They rely on the specification of equations for individual homogeneous commodities. Hence, the Japanese quality continuum would have to be segmented into a number of useful commodity units based on quality. Attempts to forecast the evolution of the Japanese beef market after deregulation have relied on this technique. The major problem has been to come up with useful single commodity units. A number of studies have used simple quality aggregates for their estimations. They include separation of beef into the following categories: imported and domestic beef; imported beef, Japanese dairy beef, and Wagyu beef; imported grass-fed beef, imported U.S. beef, Japanese dairy beef, and Wagyu beef; imported grass-fed beef, an aggregate of Japanese dairy beef and imported U.S. beef, and Wagyu beef. These aggregates, however, suppress much useful information. They do not differentiate between chilled and frozen products; they assume that all Wagyu beef is priced the same and sells into the same market segment; they assume that imported U.S. beef, all of which was frozen at the time of estimation, is interchangeable with chilled Japanese dairy beef; or that U.S. imported beef and imported grass-fed beef are interchangeable in the market. The degree of aggregation assumed simply prevents a great deal of useful information from being derived out of the estimates.

An alternative is to use data by grades to determine the interactive price effects. This provides information at a more disaggregate level and is tied directly to Japanese production. Econometric estimations, however, require long and consistent sets of data to provide the degrees of freedom required for achieving statistical reliability. Unfortunately, the Japanese introduced a new grading system on April 1, 1988. As a result, too little new information on prices and quantities by grade is available for reliable estimates to be generated. Further, imported beef is not graded and, hence, is excluded from any analysis. Given the growth of the imported sector, estimates of price-quantity interrelationships across grades is likely to be biased or derived from misspecified models.

Realistically, even if these problems could be overcome, any estimates of price-quantity interactions would be valid only under the assumption that no price decline would result from deregulation. This would mean that prices remained at approximately the same level as when the estimates were made. Even this very limited case may not be valid due to the control over import quality formerly exercised by the LIPC. Since Japanese consumers have not been allowed to express their true preferences for beef but rather to consume only a quality of beef determined by the LIPC, deregulation may lead to considerable changes in the composition of imports even in the absence of any price change.

If the price of beef does fall, the reliability of forecasts from any econometric model will likely be low. Econometric models are most accurate when forecasts are made for prices within the range of existing data. When forecasts are extrapolated beyond the range of the price data used in the estimations, statistical reliability declines very rapidly. In particular, estimates of quantities expected to be consumed at lower prices may not be valid. To use such estimates requires an assumption that consumers' behavior in response to a price change is the same at moderate or low prices as it was at very high prices. Again, this assumption is sufficiently unrealistic to seriously question any forecast.

FORMAL FORECASTING MODELS

There are two major areas of research from which forecasts of future consumption can arise. The first might generally be termed demand models or demand system models. The second could generally be described as formal forecasting models. Both types of models have econometrically estimated demand curves or systems of demand curves that relate consumption to changes in the price of beef, changes in income, and changes in the prices of substitutes and possibly some additional factors that might be expected to alter consumption. In demand or demand system models, there is no attempt to explicitly model supply aspects of the beef industry. Projections are then made on assumptions about future price and income changes (for example, the removal of the quota system leading to a fall in the price of beef imports by an expected amount). The price of imported beef is present as an argument both in its own demand equation and that of domestically produced beef in Japan. The model is "shocked" and all of the relevant interactions allowed to work through the model until new quantities of consumption and prices are generated.

Formal forecasting models, which are created for the expressed purpose of providing future projections, model both demand and supply explicitly. The models provide equilibrium levels of prices and quantities. Then the model is "shocked" by simulating the removal of a policy like the import quotas and new equilibrium prices and quantities are forecasted for each market modeled. These can be compared to the base equilibrium levels.

The signing of the 1988 Beef Market Access Agreement stimulated considerable research activity into the expected effects on Japanese consumption, particularly in major exporting countries.

Attempts to model the Japanese beef system, however, have been ongoing for some time. A study by the Australian Bureau of Agricultural and Resource Economics (1987) provided a summary of the major studies of beef demand undertaken between 1978 and 1987. Elasticities are the usual means of reporting the results of demand studies. Elasticities relate the percentage change in consumption to a percentage change in price or income. For example, an income elasticity of 1.4 for beef means that for each one percent increase in a consumer's income, the expected increase in beef consumption is 1.4 percent. While, with one exception, all the studies reported by the Australian Bureau of Agricultural Economics had similar income elasticities for beef–between 1.25 and 1.78–there was little agreement on the elasticity with respect to the change in the price of beef (the own-price elasticity). This diversity of results is understandable given the discussion above. Own-price elasticities ranged from –0.77 to –2.22, meaning that a 1 percent reduction in the price of beef would lead to an increase in beef consumption of anywhere between 0.77 and 2.22 percent. This lack of consistency makes forecasts of consumption changes due to price declines subject to such a wide range of outcomes that they are of little use. It is not surprising, however, that the range of values for estimated elasticities is so wide. All of these models treat beef in Japan as one commodity; that is, beef has a single price and different qualities are lumped together when response to price is examined. As discussed several times in this book, however, there is a price-quality continuum for beef in Japan. While it is possible to sum the quantities regardless of quality, the method of determining the price associated with each quantity is likely to be different in each study. If the single price is determined by a different method for each study, then the response estimated for a given change in price will also vary. Further, the use of such models is extremely limited for public policy because the price used to "shock" the model must be constructed in exactly the same way as the price used to estimate the model.

A number of demand studies have been undertaken with the beef market disaggregated. One such study (Mori, Lin, and Gorman, 1987) divided the market into Wagyu beef, dairy beef, and imported beef. Their work had one important conclusion: any decline in the price of imported beef was not likely to have a significant effect on either the consumption of Wagyu beef or dairy beef. This suggests that they are poor substitutes. However, the estimates were undertaken over the period when the LIPC controlled imports in such a way so as to ensure that imported beef was a poor substitute for domestically produced beef. Hence, the result should have been expected. With the removal of the LIPC from the importing system, the quality of imports will be determined by the market and is likely to change as exporters attempt to produce beef that more closely matches Japanese tastes. Hence, the demand relationships assumed in this study are no longer likely to be valid.

One major modeling effort was undertaken at the Meat Export Research Center at Iowa State University. Wahl et al. (1989) divided beef into Wagyu beef and import quality beef, which included both imported beef and Japanese dairy beef. This clearly ignores any quality differences among grain- and grass-fed imports and any quality differences between imported beef and Japanese dairy beef. The estimates yielded an own-price elasticity of −2.06 for Wagyu beef and −1.00 for import quality beef. The effect of a decreased imported beef price on the consumption of Wagyu beef was estimated to be considerable, the opposite result to the previous study. The forecast suggests that post-liberalization consumption of Wagyu beef would be 18 percent lower and dairy beef 30 percent, lower than recorded levels. Exports of U.S. beef were forecast to more than triple over their 1985 level and Australian exports to triple from their 1985 level. Van der Sluis and Hayes (1989), with an estimated own-price elasticity of −1.55 for beef in Japan, projected that beef consumption in Japan would rise from 829,000 metric tons in 1988 to 2,672,000 metric tons in 1993. Imports were projected to be 2,257,000 metric tons. The price in Japan was forecast to fall 38 percent. Wahl et al. (1988) projected that by 1997 consumption would increase to over 1,000,000 metric tons, of which about half would be imported.

Sheales (1990) reported forecasts made in Australia. Japanese beef consumption was forecast to increase from 1,160,000 metric tons in 1989 to 1,675,000 metric tons in 1995. Total imports were expected to rise from 443,000 metric tons to 780,000 metric tons over the period. Australian exports were expected to rise from 180,000 metric tons to 400,000 metric tons.

Hayes (1990c) provided estimates for 1997, based on different price and tariff regimes, that range from 709,000 metric tons to 1,106,000 metric tons for Japanese consumption. Estimates of imported grain-fed beef range from 191,000 metric tons to 595,000 metric tons and grass-fed beef from 138,000 metric tons to 224,820 metric tons.

SUMMARY

It should be apparent that there is little agreement among forecasters regarding the evolution of the Japanese beef market in response to the changes set in motion by the 1988 BMAA. This lack of consensus should also not be a surprise given the difficulties associated with modeling markets for heterogeneous goods. All models do, however, predict considerable growth. This suggests that additional resources should be expended to acquire more accurate and useful market information.

Chapter 14

Marketing to Evolving Niches

The Japanese beef market has been described as an evolving "Pandora's box" (Hobbs and Kerr, 1990). This should be of particular interest to marketers since Pandora's wonders, in this case, are a multitude of expanding markets. All too often marketing must be conducted in stable or mature markets where any increase in market share must be wrested away from existing competitors using large amounts of resources and under conditions that tax ingenuity and creativity to the limit. Competitors' resources must be countered and their strategies overcome. By contrast, the Japanese beef market presents opportunities for establishing a presence where markets are opening.

This special, although not unique, opportunity has arisen in the Japanese beef market for five fundamental reasons:

1. the Japanese economic miracle has been translated into very large increases in income, and beef consumption is very responsive to growing income;
2. the market exhibits pent-up demand for beef due to the very restrictive system of import regulations that has been in place;
3. Japanese culture is beginning to broaden its base by internalizing aspects of other cultural traditions, including those of the Western world;
4. the inability of the domestic Japanese beef industry to satisfy the growing demand due to the spatial limitations on Japanese agricultural production;
5. pressures from trading partners who are demanding increased access to Japanese markets as a condition of allowing Japanese products continued access to their markets.

The convergence of these five forces in one market at one time has led to the current opportunity. This opportunity is particularly relevant for beef producers in exporting countries due to the constraint on Japanese production.

The market should be very appealing to marketers because of its diversity. If the market for beef in Japan were relatively homogeneous, as it is in North America and Australia, the type and quality of product desired could be identified and strategies devised very quickly to satisfy the growth in demand. The market would soon reach a mature status where most marketing efforts are spent in attempting to attract customers from one's close competitors and retain one's existing customers in the face of competitors' attempts to induce them away. The Japanese market for beef appears to be sufficiently diverse that market growth can be expected on a large number of fronts. If anything, the Japanese beef market appears to be getting more diverse over time. This concept is often difficult for exporters, and particularly farmers, in exporting nations to grasp since their experience and efforts have been oriented to the production of beef defined over a very narrow range of limited quality characteristics. In other words, they have been attempting to consistently reduce the diversity implicit in biological production processes and to standardize their product.

Marketing to domestic markets in exporting countries has been largely limited to ensuring product quality, improving the efficiency of the marketing channel in the name of enhancing price competitiveness, and generic advertising aimed at reducing the market share of major meat competitors such as pork and chicken. Essentially, beef has been treated as a commodity and, hence, little product differentiation has been attempted. In North America and other parts of the Western world, where consumer tastes are relatively homogeneous, this is probably a wise course to follow. As a result, there is a dearth of persuasive marketing expertise within the beef industry itself while professional marketing agencies have little experience with the beef industry. This does not mean that product differentiation cannot be achieved in the red meat industry, as the success of the export-oriented Danish pork industry attests to, rather that it has not been really necessary to try.

Improving marketing channel efficiency has been sufficient to secure a short-run competitive advantage for beef which is similar to that provided by product differentiation in other industries. The steady movement away from butchering near the final point of sale to butchering in centralized facilities, where economies of scale can be attained, has been the major means of capturing short-run market shares in North America, Australia, and western Europe.

While the production system and marketing channels which have developed do efficiently supply homogeneous Western beef markets and have proved very adept at responding to changing preferences in these markets, they have also become a significant constraint to supplying heterogeneous foreign markets such as Japan.

The marketing system which has developed, particularly in North America, is a simple, efficient beef-marketing channel that minimizes the costs associated with geographically dispersed producers and biologically determined variability in product quality while allowing beef producers a choice of markets. This ensures a degree of competition for their product. Producers know that if their product meets certain specifications–often formally set out in a grade–they will always receive a premium price. As a result, production planning lacks a time dimension. The market is vertically segmented and this is possible because all buyers and sellers at each level require virtually the same product.

Product destined for Japan will likely differ from domestic specifications in the exporting country and, at least in the market development stage, be limited in quantity. If market niches are sufficiently narrow, this is likely to be an ongoing problem. If beef is to satisfy the better marbled portion of the Japanese quality continuum, where competition is likely to be less, then the processor must inform the packing plant, which must then inform the feedlot operator as to the type of animal to produce before feeding starts. New risks are introduced in the production and marketing of beef for this market. As a result, producers are faced with new, higher levels of risk as well as the need to acquire production information and experience. These costs may be sufficient to deter production. Devising a marketing channel that will reduce risks and provide reasonable incentives for the feedlot operator to orient practices toward products for Japan remains one of the major challenges for marketers.

This is not to suggest that the Japanese have any significant advantage in marketing expertise when it comes to beef. For all the diversity of quality that exists in the Japanese beef market, except at the highest end of the quality continuum with shimofuri and some kinds of dairy beef, there has been little attempt at positive product differentiation.

Except for some generic advertising, attempts at product differentiation have concentrated on creating a negative image of competitors, often backed by government policy. In other words, foreign beef is labeled as Australian beef or U.S. beef in an attempt to suggest that it is inferior to Japanese beef. This is a far cry from creating a positive image of the product as one that could fill a market niche. While the effect may appear to be the same, it is possible for foreign suppliers to overcome the negative connotation through product improvement. On the other hand, if a positive identification for the product is achieved, that identification is likely to remain in the mind of the consumer regardless of what the competitor does to improve or alter its product.

Most attempts at product differentiation by exporters to Japan have followed what might be termed a quasi-nationalistic approach. They have promoted their beef as Australian beef, American beef, or Canadian beef. It is not clear that the Japanese consumer really cares where the beef comes from, except for the obvious chauvinistic attachment to domestic product. Certainly, there are attempts to associate attributes that consumers value with a national image.

The primary reason a person purchases beef is in the hope that it will provide the maximum satisfaction of nutritional needs and tastes, subject to the constraints of the preparation technology available. It is the marketer's role to identify both the taste and the technology and then convince the consumer of the product's merits, using a combination of product design and positive image creation.

For that, one must return directly to the consumer. The key to marketing is consumption and what drives it. Identifying the major determinants of consumer decisions has been the reason for writing this book. The Japanese consumer is, however, elusive. That is because their tastes are neither homogeneous nor stable. Since there is no strong tradition of beef consumption in Japan, the consumer is buffeted by a large number of forces at once.

The most obvious force is the increase in income that allows for choice. Beef is one of many commodities that can be selected for consumption. There are many culinary practices and consumption norms in Japanese culture and, in part, beef has been adapting to norms which were developed to facilitate the consumption of a rice-, soya-, and fish-based cuisine. It is only very recently that consumption of beef through adaptation to traditional cuisine has been considered a mass market choice.

The ability to expand one's choice, which was provided by increased incomes, has also allowed the Japanese consumer to experiment with widening his/her culinary base through the consumption of Western foods. The cuisines of North America and Australia have a significant beef component. Their knife-and-fork technology is adapted for beef consumption. The fast-food industry has liberated the Western consumer, to some extent, from knife-and-fork technology. The success of Western-style fast-food restaurants have given the Japanese an alternative to chopsticks that is less intimidating than a knife and fork. Finger food is as an important and expanding market in Japan as it is in North America.

In short, the Japanese consumer is in a position to experiment because attitudes toward food are not fully formed. For the marketer, this means the consumer can be influenced. The winners in the race to influence the Japanese consumer's taste in beef stand to benefit from the growth potential that clearly exists. For the marketer, the important question is, Where is the exploitable market niche? Answering this has two aspects: determining the means to influence the consumer and the ability to provide the product to satisfy the consumer's desires.

Niche marketing, however, has its dangers. If the niche is too wide, or the degree of success too obvious, competitors will be attracted (in the form of emulators or close substitutes). A successful niche is one that is too small to justify a major effort by a competitor to obtain market share. If one has been successful in expanding a market niche (i.e., successfully influencing additional customers to consume the product on a regular basis), then one is likely to have an initial advantage over a potential competitor. The competitor is unlikely to be willing to expend the required resources to obtain market share if the niche is small. This does not mean that an individual company cannot exploit a number of these market niches.

However, the key remains the consumer. As yet, little work has been done on examining the Japanese consumer's selection process for beef. This is clearly the next step.

Beef consumption in Japan is extremely complex and is likely to evolve into something even more complex. As a result, successful marketing will require considerable background information so that the right questions can be asked. It was the intention of this book to provide that background. An attempt has been made to draw together, in one convenient place, a great deal of diverse information regarding the Japanese beef market and to add to this information the insights gained by the authors in the process of many years of research.

If one was looking for a cookbook on how to successfully market beef in Japan, however, the reader is by now likely to be sorely disappointed. Hopefully, some of the pitfalls to be avoided have been pointed out and the complex issues that characterize crosscultural marketing clarified. Certainly, there are sufficient examples of failures in crosscultural marketing, including some spectacular examples in the beef industry, to justify this effort. The possibilities in the Japanese beef market are certainly sufficient to justify taking the time to become informed. We hope this volume provides the means to begin.

Appendix:
The Structure of Consumer Demand

The demand for a commodity is described by the relationship between the price and the quantity of a commodity that a consumer is willing and able to purchase at a given price during a particular period of time. A consumer must be both willing and able to purchase the commodity. A consumer may have unlimited wants but scarce means by which to satisfy these wants. Thus, there is a limit to the amount of a commodity that a consumer is willing and able to purchase at each price.

Economic theory provides us with a hypothesis about demand. This is the "Law of Demand," and it postulates that the lower the price of a commodity, the larger the quantity purchased, *ceteris paribus*.[1] This law appeals to our intuition, since at higher prices less is purchased and at lower prices more is purchased.

Understanding the concept of market demand is extremely important, particularly for those attempting to supply a market with a commodity. To understand the demand for a commodity it is necessary to understand the factors that affect it. These fall into two broad groups: endogenous factors and exogenous factors.

ENDOGENOUS FACTORS

Own-Price

The most obvious effect on the quantity consumed is price. As price rises, quantity demanded falls, and as price falls, quantity demanded rises. Thus, as the price of a good increases it becomes less attractive to consumers. They may stop purchasing it altogether and switch to purchasing other goods, or they may gradually reduce the quantities they purchase. When considering the factors that might affect the demand for a product, its own price is one of the first variables to look at.

EXOGENOUS FACTORS

A number of other factors tend to affect the consumption of a commodity. Unlike changes in a commodity's own price, these exogenous factors do not represent a movement along the demand curve. Instead they represent a shift to an entirely new demand curve.

When analyzing the market demand for a commodity, several exogenous factors must be taken into consideration. These are discussed in the following sections.

Consumer Tastes

Demand for a product is obviously a function of consumer tastes. A change in tastes or preferences shifts the demand curve for a product. For example, if beef preferences in Japan changed so that consumers began to develop a taste for lean steaks, the demand curve for lean beef in Japan would shift to the right. Consumers would be willing to pay higher prices for the same quantity of beef. Or, conversely, more lean beef would be purchased at each and every price.

Advertising is an attempt to increase the demand for a product by altering tastes. For example, if an advertising campaign for beef is successful, people might switch from consuming another meat product, such as pork, to more beef instead. The demand curve for beef would then shift to the right.

The geographical location of consumers may affect their tastes for a product. For example, people in Western Japan traditionally consume more beef than those in Eastern Japan, who tend to consume more pork (Khan et al., 1990). Therefore, if one was to model the demand for beef in Japan, one might want to take into account the taste differences among various regions or over time.

Population

An increase in population size, *ceteris paribus*, increases the total demand for a product. Thus, one might expect the demand for foodstuffs to increase in a country with a rapidly growing population. However, this must be an effective demand; that is, for demand to increase requires not only a population increase, but that this increased population has the ability (i.e., the funds) to increase its demand.

The second way in which population can affect demand for a commodity is through a change in the population's demographics. If, for example, older people constitute a larger proportion of the population then there will be an increase in the types of commodities demanded by this segment.

This could occur in Japan since the percentage of the population over the age of 65 has been growing in recent years (Khan et al., 1990). If it is found, for example, that older segments of the population prefer traditional Japanese dishes over modern North-American-style fast-food dishes, then the demand for traditional Japanese dishes will increase as the proportion of the population over 65 increases.

The geographical distribution of the population may also affect demand for a product. Thus, migration of the population from rural to urban areas will increase the overall demand for the types of products demanded by urban dwellers and decrease the overall demand for the types of products demanded by rural dwellers. For example, the urban lifestyle may be more hectic than that in rural areas, and this may lead to an increase in the demand for meals consumed away from home. The demand curve for restaurant meals would then shift to the right.

Thus, changes to both the size and the social and regional distribution of the population have to be taken into account when studying the consumption of a product.

Income

Personal disposable income is an extremely important determinant of the demand for a commodity. As incomes rise, one would expect to see demand increase for a normal good, *ceteris paribus*.

As income falls, one would expect to see a fall in the demand for a normal good *ceteris paribus*. A normal good is one for which demand increases as income increases. Beef is generally considered to be a normal good.

There are some goods that do not obey this income law. These are known as "inferior" goods. The consumption of an inferior good falls as income rises. An example of an inferior good is an inexpensive basic foodstuff, such as potatoes in a poor country. As incomes rise, people become able to afford more tasty or nutritious foods, such as red meats, and switch from consuming the basic foodstuff.

For the majority of commodities, and certainly for beef consumption in Japan, it is assumed that the good is normal. An increase in incomes, *ceteris paribus*, could be expected to result in an increase in the demand for beef, shifting the demand curve outwards. There may be, however, some beef products considered to be of sufficiently low quality such that they will exhibit declining consumption as incomes rise.

Government Regulations

Government-imposed regulations may also have an impact on the effective demand for a commodity.

For example, legislation prohibiting alcohol consumption in public places may reduce the demand for alcoholic beverages. People will be forced to buy soft drinks rather than alcoholic drinks when in a public place. Thus, less beer would be demanded at each and every price, and the demand curve for beer would shift to the left. Similarly, health and sanitation regulations might restrict the types of products consumers can purchase.

While government-imposed restrictions on demand are expected to reflect the desires of consumers over the long run, they will always distort the consumption patterns of some individuals unless there is a consensus among all consumers that concurs with the regulation. Of course, if there was such a consensus, there would be no need for the regulation. Regulations do, however, affect the demand for a product and, where they are in place, their effect should be taken into account. If the effects of government regulations are not fully understood then the effective demand for a product in a market may be over- or underestimated.

The Price of Related Commodities

Related commodities can affect consumption of a good in either a positive or a negative way. If a decrease in the price of a related second commodity decreases the quantity consumed of the first good, the two goods are called *substitutes*. If a decrease in the price of a related second commodity increases the quantity consumed of the first good, the two goods are called *complements*.

Substitutes are those commodities that satisfy similar needs or desires as the product in question. Thus, a substitute for beef might be another meat product such as pork, lamb, or chicken. If the price of this substitute good falls, it will negatively affect demand for beef. Conversely, if the price of pork rises, *ceteris paribus*, we would expect to see a rise in the demand for beef–more demanded at each and every price since beef would be relatively cheaper compared with pork.

A complementary good is one that is used jointly with another good. Steak sauce is an example of a complementary good for beef. If the price of steak sauce decreases, the demand for beef will increase, *ceteris paribus*. The demand curve for beef will shift to the right; more is consumed at each and every price. Likewise, if the price of steak sauce rises, *ceteris paribus*, we might expect to see a fall in the demand for beef. The demand curve will shift to the left; less will be consumed at each and every price.

Other Factors

The above discussion lists the main factors that influence any product's consumption. However, other factors may affect demand, which are particular to the product in question.

Demand for some commodities is affected by events that occur within the context of the market, for example, changes in weather or changes in season. The demand for beef in North America has a seasonal pattern, with more beef demanded in the summer months when people tend to barbecue more often. The demand for poultry, particularly turkey, *ceteris paribus*, increases during Christmas and Thanksgiving periods. The demand for heating oil depends, to some extent, on average winter temperatures.

These factors act to shift the demand curve for the product in question, just as all the other exogenous factors mentioned above shift the demand curve. When analyzing product consumption it is necessary to include all the factors mentioned above as well as any additional factors that might help explain demand in that particular market.

SUMMARY

The demand for a product is affected by a number of factors. Demand is inversely related to own price, so that as price increases, quantity demanded falls (and vice versa). These are movements along a given demand curve. All other factors affect demand by shifting the demand curve so that a different quantity is consumed at each and every price. These are exogenous factors, the most important of which are consumer tastes, population, income, government regulations, and the prices of related products.

Notes

Chapter 1

1. The Japanese Fiscal Year runs from April 1 to March 31. For example, Japanese Fiscal Year 1992 would begin on April 1, 1992, and end on March 31, 1993.

2. Longworth (1983) and the Australian Bureau of Agricultural and Resource Economics (1987) Chapter 7 provides more detailed discussions of the special quotas and the JMC and LIPC levies.

3. Chilled is the meat industry term for fresh. Chilled beef is normally kept at a temperature of approximately 0 degrees Celsius.

4. Thus, the LIPC hampered the normal vertical coordination mechanisms of the market. Vertical coordination describes the ways in which a product moves through the production and distribution system.

5. There were two price bands: one for Wagyu cattle, the Japanese beef breed which produces the high quality marbled beef; and one for dairy cattle, the source of lower quality beef in Japan. The Wagyu price band is at higher price levels than the dairy price band. Since the price stabilization scheme was introduced, neither price approached the floor of its band, yet both prices have at times exceeded the ceilings of the bands. In later years, a single price band was set.

6. A bound tariff means that the tariff cannot be raised above that level in the future.

7. The import tariff will be calculated on the basis of the value of the beef plus its transportation cost. Air freighting of beef to Japan can be seven times more expensive than ocean freighting beef (Hobbs, 1990). Thus, the tariff may make air freighting of beef prohibitively expensive.

Chapter 2

1. The Japanese beef-grading system is not discussed in any detail here. A new system was introduced in April 1988 which has three letter grades, from A to C, with A as the highest; these measure the meat yield. The grades are calculated so that it is almost impossible for non-Wagyu beef to achieve an A grade. The second part of the grading system is a number grade for meat quality. This measures such things as marbling, fat color, and tenderness. There are five number grades with 5 as the highest. Thus, there are 15 possible Japanese beef grades ranging from A5 to C1. The highest prices are paid for grade A5 beef.

2. Fermented bean paste, which can be red or white, sweet or salty.

3. Boiled rice and other food flavored with vinegar. The most popular form of Sushi is rice cakes plastered with raw fish or rolled in seaweed and sprinkled with vinegar.

Chapter 7

1. Full sets are boxed beef products in which all of the main primals from a carcass are placed in the box. It can be thought of as a dressed carcass that has been divided and packed in a box for shipping.

Chapter 9

1. For example, if a beef exporter expected to receive Y600 per kilogram at an exchange rate of U.S. $1.00 = Y120, the expected return would be U.S. $5.00 per kg. If the exporter's price expectations prove to be correct, when the cattle are eventually sold, they will earn Y600/kg in the Japanese market. However, if over the same period, the Japanese Yen were to depreciate in value against the dollar to a level of U.S. $1.00 = Y150, the exporter would now only receive $4/kg for the beef.

2. Other mechanisms are, of course, feasible. These include forward markets, risk pooling, and a futures market in well-marbled beef. These have been discussed elsewhere [see Kerr et al. (1990b) and Hayes (1989)].

3. Although it has been argued that a total feeding period of at least 300 days for the Japanese beef market would be desirable, little is known about the performance of non-Japanese breeds of cattle fed for this length of time. Feed conversion ratios and daily weight gains cannot be predicted with any certainty until actual feed trials have been carried out. Previous analyses of the Japanese beef market have used a hypothetical feeding period of 260 days [see Lin and Mori (1990) and Hobbs (1990)].

4. The risks inherent in contractual agreements, and particularly the problem of PCOB, have been described by Teece (1976):

> Even when all of the relevant contingencies can be specified in a contract, contracts are still open to serious risks since they are not always honored. The 1970s are replete with examples of the risks associated with relying on contracts . . . open displays of opportunism are not infrequent and very often litigation turns out to be costly and ineffectual. (p. 5)

It is this risk, the failure of one party to uphold a contract, that is likely to make the use of forward contracts unreliable in the market for heavily marbled beef.

5. Since the tariff is calculated on a c.i.f. basis, the internal accounts of the firm could "undervalue" the product, taking account of additional costs once the product is inside Japan and the import tariff has been paid.

Chapter 10

1. This consumer survey was conducted by Japan Meat Service and Information Center.

2. The more important of these are the following: Marketing Intelligence Corporation; the Institute of Marketing Research and Statistics; Marketing Research Service; Audience Studies (Japan) Inc. (ASI); A.C. Nielson Company (Japan); Dentsu Research Ltd.; Marketing Center; Japan Marketing Research; Central Research Services Inc.; International Marketing Services; Japan Market Research Bureau; and Fuji National Consulting Corporation (De Mente, 1988).

3. The same can be said for Australian, Canadian, or any other source of imported beef.

Chapter 11

1. Thus, any gain in revenue from the increased demand for imported beef that follows a price decline will be outweighed by the loss in revenue from selling each unit at a lower price.

2. That is, Wagyu heifer grades A5 to B1, Wagyu steer grades A5 to B1, Dairy heifer grades B5 to C1 and Dairy steers grades B5 to C1. The grading system is constructed such that only Wagyu cattle are able to attain the A range of grades.

Appendix

1. *Ceteris paribus* is a latin phrase meaning "all other things held constant." It allows us to consider what happens to the demand for a commodity after a change in one variable, such as price, while all other variables that might affect demand are assumed to be unchanged.

References

Armington, P.S. 1969. "A Theory of Demand for Products Distinguished by Place of Production." *International Monetary Fund Staff Papers,* 16: 159-199.

Australian Bureau of Agricultural and Resource Economics. 1987. *Japanese Beef Policies: Implications for Trade Prices and Market Shares.* Canberra: Australian Government Publishing Service.

Boutin, B.D. and Kerr, W.A. 1990. "The Importance of the Red Meat Industry to the Canadian Economy." *Meat Probe,* 6 (2): 1-3.

Canada, Statistics Canada. *Food Consumption Statistics.* Ottawa.

Clarke, R. 1989. *Food and Beverage Trends in Japan.* Tokyo: ASI Market Research Inc.

Colman, D. and Miah, H. 1973. "On Some Estimates of Price Flexibilities and Their Interpretation." *Journal of Agricultural Economics,* 24: 353-367.

Coyle, W.T. 1983. *Japan's Feed-Livestock Economy.* Washington: Foreign Agricultural Economic Report No. 177, Economic Research Service, USDA.

Coyle, W.T. 1986. *The 1984 U.S.-Japan Beef and Citrus Understanding: An Evaluation.* Washington: Foreign Agricultural Economic Report No. 222, Economic Research Service, USDA.

De Mente, B. 1988. *How to do Business with the Japanese: A Complete Guide to Japanese Customs and Business Practices.* Lincolnwood: NTC Business Books.

Dentsu Japan. 1983. *Marketing/Advertising Yearbook, 1983.*

Dentsu Japan. 1989. *Marketing/Advertising Yearbook, 1989.*

Department of External Affairs. 1988. *Export Opportunities in Japan–The Food Service Market.* Ottawa: Government of Canada.

Fielding, R. 1990. "Constraints on a New Exporter–Domestic Regulations and Market Access." In *Selling Beef to Japan,* edited by Elton, D.K., Kerr, W.A., Klein, K.K. and Penner, E.T. Calgary: Canada West Foundation, pgs. 145-152.

Gomez-Casseres, B. 1987. "Joint Venture Instability: Is It a Problem?" *Columbia Journal of World Business,* 22: 97-102.

Gorman, W.D. 1990. "Beef in Japan in the 1990's: The Players, the Rules and the Payoffs." In *Selling Beef to Japan,* edited by Elton, D.K., Kerr, W.A., Klein, K.K. and Penner, E.T. Calgary: Canada West Foundation, pgs. 17-30.

Harrison, G. 1990. "Opportunities and Strategies for Beef Exports to Japan: New Zealand." In *Selling Beef to Japan,* edited by Elton, D.K., Kerr, W.A., Klein, K.K. and Penner, E.T. Calgary: Canada West Foundation, pgs. 81-92.

Hayami, H. and Yamada, S. 1991. *The Agricultural Development of Japan: A Century's Perspective.* Tokyo: University of Tokyo Press.

Hayes, D.J. 1989. *Beefing Up for the Japanese Market. Trade Issues and Opportunities.* Des Moines and Ames: Paper No. 1, Midwest Agribusiness Trade Research and Information Center.

Hayes, D.J. 1990a. *The Economics of Feeding, Processing and Marketing Beef Animals for Export to the Pacific Rim.* Presented at the 1990 Annual Agricultural Economics and Farm Management Society, Vancouver, British Columbia, August 5-8.

Hayes, D.J. (ed.) 1990b. *Meat Marketing in Japan: A Guide for U.S. Meat Exporting Companies.* Des Moines and Ames: Midwest Agribusiness Trade Research and Information Center.

Hayes, D.J. 1990c. "Opportunities and Strategies for Beef Exports to Japan: A U.S. Perspective." In *Selling Beef to Japan,* edited by Elton, D.K., Kerr, W.A., Klein, K.K. and Penner, E.T. Calgary: Canada West Foundation, pgs. 33-63.

Hayes, D.J., Green, J.R. and Wahl, T.I. 1990. "Meat Marketing in Japan: Economic Considerations." In *Meat Marketing in Japan: A Guide for U.S. Meat Exporting Companies,* edited by Hayes, D.J. Des Moines and Ames: Midwest Agribusiness Trade Research and Information Center, pgs. 159-228.

Hobbs, J.E. 1990. *An Analysis of the Opportunities, Risk and Constraints Facing the Canadian Beef Industry Following the Liberalization of Japanese Beef Import Regulations.* Calgary: Unpublished M.A. thesis, University of Calgary.

Hobbs, J.E. and Kerr, W.A. 1990. "Changing Pandora's Box: Liberalization of the Japanese Beef Importing System." In *Selling*

Beef to Japan, edited by Elton, D.K., Kerr, W.A., Klein, K.K. and Penner, E.T. Calgary: Canada West Foundation, pgs. 5-15.

International Monetary Fund (IMF). 1991. *International Financial Statistics.* Washington, D.C.

Japan Meat Service and Information Center. 1988. *Seasonal Report on Consumer Movement for Meat Consumption.* Tokyo.

Japan Ministry of Agriculture, Forestry and Fisheries (Japan MAFF). 1992. *The Meat Statistics in Japan.* Tokyo: Japan Ministry of Agriculture, Forestry and Fisheries.

Japan Statistics Bureau. 1988. *Japan Statistical Yearbook.* Tokyo: Management and Coordination.

Kerr, W.A. 1985. "The Livestock Industry and Canadian Economic Development." *Canadian Journal of Agricultural Economics,* 32: 64-104.

Kerr, W.A. 1987. "The Recent Findings of the Canadian Import Tribunal Regarding Beef Originating in the European Economic Community." *Journal of World Trade Law,* 21 (5): 55-65.

Kerr, W.A. and Cullen, S. 1990. "A Marketing Strategy for Canadian Beef Exports to Japan." In *Canadian Agricultural Trade: Disputes, Actions and Prospects*, edited by Lermer, G. and Klein, K.K. Calgary: University of Calgary Press, pgs. 209-230.

Kerr, W.A. and Hobbs, J.E. 1989. *The Canadian Perspective on Japanese Beef Import Liberalization.* Presented to the Annual Meeting of the Technical Committee of the Western Region Project (W-177)–Domestic and International Marketing Strategies for U.S. Beef, U.S. Department of Agriculture, Honolulu, November 30-December 1.

Kerr, W.A., Hobbs, J.E. and Gillis, K.G. 1990a. "Reducing the Risk of Exporting Beef to Japan: Development of Market Information." In *Selling Beef to Japan*, edited by Elton, D.K., Kerr, W.A., Klein, K.K. and Penner, E.T. Calgary: Canada West Foundation, pgs. 171-190.

Kerr, W.A., Hobbs, J.E. and Gillis, K.G. 1990b. "Managing the Risk of Dealing with Countries in the Pacific Rim." *Canadian Journal of Agricultural Economics,* 38: 911-921.

Kerr, W.A. and Klein, K.K. 1989. *Western Canadian Beef in the Japanese Market: Opportunities and Challenges.* Calgary: Canada West Foundation.

Khan, L., Ramaswami, S. and Sapp, S.G. 1990. "Meat Marketing in Japan." In *Meat Marketing in Japan: A Guide for U.S. Meat Exporting Companies*, edited by Hayes, D. Des Moines and Ames: Midwest Agribusiness Trade Research and Information Center, pgs. 9-77.

Klein, B., Crawford, R.G. and Alchian, A.A. 1978. "Vertical Integration, Appropriable Rents and the Competitive Contracting Process." *Journal of Law and Economics,* 21: 297-326.

Klein, P.L., Klein, K.K. and Yoshida, S. 1990. "Attitudes Towards Beef in Hokkaido, Japan." *Kitami Daigaku Ronshu,* 23: 41-55.

Knipe, C.L. and Rust, R.E. 1989. *Fresh and Processed Pork for the Japanese Market.* Ames: MERC Staff Report #7-86, Meat Export Research Center.

Kohls, R.L. and Uhl, J.N. 1990. *Marketing of Agricultural Products.* 7th Edition. New York: Macmillan Publishing Company.

Lin, B. and Mori, H. 1990. *Niku: A Japanese Beef Market Analysis Program, Version 1.0.* Moscow: Department of Agricultural Economics, University of Idaho, mimeo.

Lin, B., Mori H., Gorman, W.D. and Rimbey, N.R. 1989. *Producing Beef for the Japanese Market.* Moscow: Regional Project R894, Agriculture Experiment Station, University of Idaho.

Lloyd, R., Frank, M.D. and Faminow, M.D. 1987. *Japanese Beef Trade: Some Political and Economic Considerations.* Tucson: Report No. 38, Department of Agricultural Economics, University of Arizona.

Longworth, J.W. 1983. *Beef in Japan: Politics, Production, Marketing and Trade.* St. Lucia: University of Queensland Press.

Martin, L.G. 1989. The Graying of Japan. *Population Bulletin,* 44 (3): 4-39.

Mighell, R.L. and Jones, L.A. 1963. *Vertical Coordination in Agriculture.* Washington, DC, USDA ERS-19. As reported on p. 760, in Lang, M.C. 1980. "Marketing Alternatives and Resource Allocation: Case Studies of Collective Bargaining." *American Journal of Agricultural Economics,* 62: 760-765.

Mori, H. 1990. Comments made during "Question Period–Session II." In *Selling Beef to Japan,* edited by Elton, D.K., Kerr, W.A., Klein, K.K. and Penner, E.T. Calgary: Canada West Foundation, pgs. 159-167.

Mori, H. and Lin, B. 1990. "Institutional Constraints to the Import and Distribution of Beef in Japan." in *Selling Beef to Japan,* edited by Elton, D.K., Kerr, W.A., Klein, K.K. and Penner, E.T. Calgary: Canada West Foundation, pgs. 123-132.

Mori, H., Lin, B. and Gorman, W.D. 1987. *Measuring Demand Functions for Imported Beef in the Japanese Market.* Las Cruces: Department of Agricultural Economics, New Mexico State University.

Nakase, S. 1990. "The Recent Beef Situation." *Japanese Beef Industry Facing Trade Liberalization.* Report Study Group on International Issues SG11 No. 6. Japan: Bulletin of the Food and Agriculture Policy Research Center, pps. 109-174.

Namiki, M. 1990. "The Trend in Beef Prices Before Overall Liberalization." *Japanese Beef Industry Facing Liberalization.* Report Study Group on International Issues SG11 No. 6. Japan: Bulletin of the Food and Agriculture Policy Research Center, pgs. 41-107.

O'Connell, J. 1986. "A Hedonic Price Model of the Paris Lamb Market." *European Review of Agricultural Economics,* 13: 439-450.

Ogura, T. 1983. "Food Policy Study." No. 2, Vol. 34, as reported in Namiki, M. 1990. "The Trend in Beef Prices Before Overall Liberalization." *Japanese Beef Industry Facing Trade Liberalization.* Report Study Group on International Issues SG11 No. 6. Japan: Bulletin of the Food and Agriculture Policy Research Center, pgs. 41-107.

Organization of Economic Cooperation and Development (OECD). 1986. *Meat Balances in OECD Countries 1980-1986.* Paris.

Organization of Economic Cooperation and Development (OECD). 1990. *OECD Outlook,* 47, Paris.

Organization of Economic Cooperation and Development (OECD). 1991. *Meat Balances in OECD Countries. 1983-1989.* Paris.

Organization of Economic Cooperation and Development (OECD). 1992a. *Agricultural Policies, Markets and Trade.* Paris.

Organization of Economic Cooperation and Development (OECD). 1992b. *Main Economic Indicators.* Paris.

Owen, D. 1989. "Your Beef Men Need Us, Say Japanese." *The Weekend Australian,* January 14-15: 3.

Perdikis, N. and Hobbs, J.E. 1990. "Opportunities and Strategies for Beef Exports to Japan: The European Community." In *Selling Beef to Japan*, edited by Elton, D.K., Kerr, W.A., Klein, K.K., and Penner, E.T. Calgary: Canada West Foundation, pgs. 93-113.

Savell, J.W. 1990. "Extending the Shelf-life." In *Selling Beef to Japan*, edited by Elton, D.K., Kerr, W.A., Klein, K.K. and Penner, E.T. Calgary: Canada West Foundation, pgs. 193-196.

Seim, E.L. 1990. "Logistics Concerns for Meat Transportation to Japan from the United States." In *Meat Marketing in Japan: A Guide for U.S. Meat Exporting Companies*, edited by Hayes, D. Des Moines and Ames: Midwest Agribusiness Trade Research and Information Center, pgs. 113-130.

Sheales, T. 1990. "Opportunities and Strategies for Beef Exports to Japan: The Australian View." In *Selling Beef to Japan*, edited by Elton, D.K., Kerr, W.A., Klein, K.K. and Penner, E.T. Calgary: Canada West Foundation, pgs. 65-80.

Simpson, J.R., Yoshida, T., Miyazaki, A. and Kada, R. 1985. *Technological Change in Japan's Beef Industry*. Boulder: Westview Press.

Stelfox, D. 1990. "Experience of a Successful Exporter of Beef to Japan." In *Selling Beef to Japan*, edited by Elton, D.K., Kerr, W.A., Klein, K.K. and Penner, E.T. Calgary: Canada West Foundation, pgs. 153-157.

Stent, W.R. 1967. "An Analysis of the Price of British Beef." *Journal of Agricultural Economics,* 28: 121-131.

Takahashi, I. 1990. "Trade Liberalization of Beef in Japan–Its Features and Prospective Impacts." *Japanese Beef Industry Facing Liberalization.* Report of Study Group on International Issues, No. 6. Tokyo: Food and Policy Research Centre Japan, pgs. 1-40.

Teece, D.J. 1976. *Vertical Integration and Divestiture in the U.S. Oil Industry,* 31. As reported in Klein, B., Crawford, R.G. and Alchian, A.A. 1978. "Vertical Integration, Appropriable Rents and the Competitive Contracting Process." *Journal of Law and Economics,* 21: 297-326.

Van der Sluis, E. and Hayes, D.J. 1989. *An Assessment of the 1988 Japanese Beef Market Access Agreement on Beef and Feed-Grain Markets.* Paper presented at the 1989 annual meeting of the American Agricultural Economics Association, Baton Rouge.

Wahl, T.I., Hayes, D.J. and Williams, G.W. 1987. *Japanese Beef Policy and GATT Negotiations: An Analysis of Reducing Assistance to Beef Producers.* Working Paper #87-7, International Agricultural Trade Research Consortium, presented at the Pacific Economic Cooperation Conference. Napier, New Zealand, October 19-22.

Wahl, T.I., Hayes, D.J. and Williams, G.W. 1988. *Dynamic Adjustment in the Japanese Beef Industry Under Alternative Import Policy Regimes.* Journal Paper No. J-13379. Ames: Iowa Agriculture and Home Economics Station.

Wahl, T.I., Williams, G.W. and Hayes, D.J. 1989. "1988 Japanese Beef Market Access Agreement: Forecast Simulation Analysis." In *Government Intervention in Agriculture,* edited by Greenshields, B. and Bellamy, M. IAAE Occasional Paper No. 5, 1989.

Wright, R.W. 1979. "Joint Venture Problems in Japan." *Columbia Journal of World Business,* 14 (1): 25-31.

Yoshida, S. and Klein, K.K. 1990. "Constraints on Beef in the Japanese Diet: Culture, Culinary Arts and Quality." In *Selling Beef to Japan,* edited by Elton, D.K., Kerr, W.A., Klein, K.K. and Penner, E.T. Calgary: Canada West Foundation, pgs. 133-143.

Index